Mechanical Technical Interview

All Important Mechanical Engineering Technical Interview Questions & Answers covering all the subjects, Important for Viva Exams & Job Interviews for Freshers and Experienced.

This book has been written by keeping in mind of various competitive exams and interviews of all kind of organizations. This book caters to the syllabus of almost all Universities and all the topics of Mechanical Engineering.

Contents:

(Abbreviation used: SOM - Strength of Materials, KOM - Kinematics of Machines)

Design, SOM, KOM, Manufacturing Process, Engineering Mechanics & Vibrations

01. What is mechanical engineering?

Answer: Mechanical engineering is the discipline that applies the principles of engineering, physics and materials science for the design, analysis, manufacturing and maintenance of mechanical systems. It is the branch of engineering that involves the design, production and operation of machinery and tools.

02. What is the difference between Technology and Engineering?

Answer: Answer: Engineering is application of science. Technology shows various methods of Engineering. A bridge can be made by using beams to bear the load, by an arc or by hanging in a cable; all shows different technology but comes under civil engineering and science applied is laws of force/load distribution.

03. What is the difference between Speed and Economic Speed?

Answer: The rated speed tells us about the maximum speed which can be achieved by a vehicle or some other machine but the economical speed means the speed limit at which the machine works efficiently with least consumption of fuel. e.g. in normal bikes(not racing),the maximum speed limit shown on speedometer is up to 120 kmph but companies always advice their customers to drive such bikes at around 60 kmph to have maximum mileage.

04. What causes hardness in steel? How heat treatment alters properties of steel?

Answer: The shape and distribution of the carbides in the iron determines the hardness of the steel. Carbides can be dissolved in austenite is the basis of the heat treatment of steel. If steel is heated above the A critical temperature to dissolve all the carbides, and then cooled, suitable cooling through the cooling range will produce the desired size and distribution of carbides in the ferrite, imparting different properties.

05. What is the difference between isotropic and anisotropic materials?

Answer: If a material exhibits same mechanical properties regardless of loading direction, it is isotropic, e.g., homogeneous cast materials. Materials lacking this property are anisotropic.

06. What are orthotropic materials?

Answer: It is a special class of anisotropic materials which can be described by giving their properties in three perpendicular directions e.g. wood; composites.

07. What is view factor?

Answer: View factor is dependent upon geometry of the two surfaces exchanging radiation.

08. Explain difference between fissile and fertile materials.
Answer: The materials which can give nuclear fission e.g. U 35, Pu 39, U 33 are fissile materials. Fertile material itself is not fissionable, but it can be converted to a fissionable material by irradiation of neutrons in a nuclear reactor.

09. Mention two types of dislocations.
Answer: Dislocation refers to a break in the continuity of the lattice. In edge dislocation, one plane of atoms gets squeezed out. In screw dislocation the lattice atoms move from their regular ideal positions.

10. What is Powder Technology?
Answer: Powder technology is one of the ways of making bearing material. In this method metals like bronze, Al, Fe are mixed and compressed to make an alloy.

11. What are the principal constituents of brass?
Answer: Principal constituents of brass are copper and zinc.

12. What is Curie point?
Answer: Curie point is the temperature at which ferromagnetic materials can no longer be magnetised by outside forces.

13. By which instruments the shear stress in fluids can be measured directly?
Answer: By Stanton tube or Preston tube.

14. Explain the difference between the points of inflexion and contraflexure.
Answer: At points of inflexion in a loaded beam the bending moment is zero and at points of contraflexure in loaded beam the bending moment changes sign from increasing to decreasing.

15. What is the difference between proof resilience and modulus of resilience?
Answer: Proof resilience is the maximum strain energy that can be stored in a material without permanent deformation. Modulus of resilience is the maximum strain energy stored in a material per unit volume.

16. What do you understand by critical points in iron, iron-carbide diagram?
Answer: The temperatures at which the phase changes occur are called critical points (or temperatures).

17. Define buckling factor.
Answer: It is the ratio of the equivalent length of column to the minimum radius of gyration.

18. Explain Bicycle Rear Wheel Sprocket working?
Answer: Rear wheel sprocket works under the principle of ratchet and pawl.

19. How to Find, Ductile-Brittle Transition Temperature in Metals?

Answer: The point at which the fracture energy passes below a pre-determined point for a standard Impact tests. DBTT is important since, once a material is cooled below the DBTT, it has a much greater tendency to shatter on impact instead of bending or deforming.

20. What is the difference between P11 and P12 Pipes?

Answer: P11 the chromium molybdenum composition that is 1% of chromium and 1/4% of molybdenum P12 the chromium molybdenum composition that is 1% of chromium and 2% of molybdenum.

21. State the difference between Unilateral and Bilateral Tolerance?

Answer: A unilateral tolerance is tolerance in which variation is permitted only in one direction from the specified direction. e.g. 1800 +0.000/-0.060

Bilateral tolerance is tolerance in which variation is permitted in both directions from the specified direction. e.g. 1800 +0.060/-0.060

22. What is the abbreviation of welding rod 7018?

Answer: 7018 is equal to

        70 = tensile strength 70000psi

        1= welding position

        8 = current flux

23. State the difference between Anti Friction Bearing and Journal Bearing.

Answer: Generally, journal bearings have higher friction force, consume higher energy and release more heat, but they have larger contact surface, so normally used in low speed high load applications. In anti friction bearings friction is less. One object just rolls over each other.

24. What is difference between Welding and Brazing?

Answer: In Welding concentrated heat (high temperature) is applied at the joint of metal and fuse together.

In Brazing involves significantly lower temperatures and does not entail the melting of base metals. Instead, a filler metal is melted and forced to flow into the joint through capillary action.

25. How to calculate bearing number from shaft Diameter?

Answer: Divide the shaft diameter size by 5, it will give last two digits of the bearing no. and according to type of load we have to choose the type of bearing and that will give prior number of the bearing.

26. The Fatigue life of a part can be improved by?

Answer: Improving the surface finish by Polishing & providing residual stress by Shot peening.

27. Poisson's Ratio is higher in Rubber, Steel or Wood?

Answer: When a material is compressed in one direction, it usually tends to expand in the other two directions perpendicular to the direction of compression. This phenomenon is called the Poisson effect. Poisson's ratio is a measure of the Poisson effect.

     For rubber = 0.5

     For steel = 0.288

     For wood < 0.2

Thus Poisson's ratio is higher in RUBBER.

28. What is the other name of Micrometer & Vernier Calliper?

Answer: Micrometer's other name is Screw Gauze & Vernier calliper's other name is slide calliper.

29. What is the need for drafting?

Answer: Drafting is the allowances give to casting process. It also used to remove the casting from mould without damage of corners.

30. What is the difference between BSP thread and BSW thread?

Answer: The British Standard Pipe thread (BSP thread) is a family of standard screw thread types that has been adopted internationally for interconnecting and sealing pipe ends by mating an external (male) with an internal (female) thread.

British Standard Whitworth (BSW) is one of a number of imperial unit based screw thread standards which use the same bolt heads and nut hexagonal sizes.

31. What is the amount of carbon present in Cast Iron?

Answer: Carbon is basically present in the form of cementite in cast iron. Its percentage lies in the range of 2.03-6.67% by weight of cementite for Cast Iron. If the amount is less than the above range than it is stainless steel

32. What are the loads considered when designing the Nut and Bolts?

Answer: Shear Loads & crushing loads

33. What is the difference between a Fence and a Wall?

Answer: A fence is either more temporary or constructed from materials, other than concrete, stone or brick.

34. What is the Difference between Quantitative and Qualitative Research?

Answer: Quantitative research involves gathering data that is absolute, such as numerical data, so that it can be examined in as unbiased a manner as possible.

Qualitative research may yield stories, or pictures, or descriptions of feelings and emotions. The interpretations given by research subjects are given weight in qualitative research, so there is no seeking to limit their bias. At the same time, researchers tend to become more emotionally attached to qualitative research, and so their own bias may also play heavily into the results.

35. What is Bending moment?
Answer: When a moment is applied to bend an element, a bending moment exists in the element.

36. What are the points in the Stress Strain curve for Steel?
Answer: Proportional limit, elastic limit or yield point, ultimate stress and stress at failure.

37. Define Reynolds number.
Answer: Reynolds number is the ratio of inertial force and viscous force. It is a dimensionless number. It determines the type of fluid flow.

38. How many Joules is 1 BTU? What is PS?
Answer:
1 BTU is equal to 1055.056 joules.
PS is Pferdestarke, the German unit for Horsepower.

39. Explain the nomenclature of a 6203-ZZ bearing.
Answer: 6 is the type code, which shows it is a single-row ball bearing, 2 is the series, means light, 03 is the bore, which is 17 mm and ZZ is the suffix meaning double shielded bearing.

40. What is Gear ratio?
Answer: It is the ratio of the number of revolutions of the pinion gear to one revolution of the idler gear.

41. What is Annealing?
Answer: It is a process of heating a material above the re-crystallization temperature and cooling after a specific time interval. This increases the hardness and strength if the material.

42. Define Torque.
Answer: Torque is defined as a force applied to an object those results in rotational motion.

43. What is Ductile-Brittle Transition Temperature?
Answer: It is the temperature below which the tendency of a material to fracture increases rather than forming. Below this temperature the material loses its ductility. It is also called Nil Ductility Temperature.

44. What is the difference between Pipe and Tube?
Answer: Ex. Iron Pipe, Carbon Tube, Steel Tube etc.

Tube is defined by Outer diameter and Wall thickness (OD and WT). While Pipe is defined by Inner diameter (ID).

Example:
   i.   Inch pipe have 2.375 inch outer diameter, where 2 inch tube have same 2 inch outer diameter.
  ii.   Tube is easily shaped by bending, where Pipe needs some mechanical work to shape it.
 iii.   Tube is tight then Pipe.
  iv.   Tube is applicable to non cylindrical sections like Square and Rectangular.

45. How does Welding damage Eye sight?

Answer: An Electric welding arc produces Ultraviolet light and the UV light (Sun burn) will damage the retina. Welding shields or Goggles with the proper shade (Automatic shading) of lens is the best protection for welders. Light filtering curtains and reduced reflective surfaces help protect both welders and observers in the area.

46. Difference between Codes, Standards and Specifications

Answer:
   1.   Code is procedure of acceptance and rejection criteria.
   2.   Standard is accepted values and compare other with it.
   3.   Specification is describing properties of any type of materials.

47. What is Auto Dosing?

Answer: Auto dosing is an automated system of feeding the equipment with liquid products. It is the ideal way to ensure the correct calibrated dose at the right time every time in auto.

48. Which is heavier 1kg of Iron or 1kg of Cotton? And why?

Answer: Both of them have same weight. The only difference is the volume of Iron is small compared to Cotton.

49. Explain why BCC, FCC and CPH lattice packing and features of grain structure affect the Ductility and Brittleness of parent metallic materials?

Answer: Ductility is the mechanical property of a material. (It is the material's ability to deform under the tensile stress without fracture). So it is depends on the atoms how they arranged in a lattice and its grain size. The ability to absorb the energy of the impact and fracture resistance depends on the arrangement of the atoms in a lattice and features of grain structure.

50. Why do you have Truss Bridges?

Answer: Truss bridges carry load in tension and compression rather than bending (Self weight + the weight of vehicles crossing it + Wind loads). A truss has the ability to dissipate a load through the truss work. The beams are usually arranged in a repeated triangular pattern, since a triangle cannot be distorted by stress.

51. Why I-section beam is preferred for heavy loading?

Answer: Cross sectional shape I, giving many benefits. It is very good for giving stiffness (less deformation on loading) and to withstand higher bending moments (as a result of heavy loading) on comparison with other cross-sectional shapes of same area. Also, it is very easy to manufacture. It will have more moment of inertia.

52. What is difference between Center of Mass and Center of Gravity?

Answer: Both terms are same when gravity is uniform. When gravity is non-uniform following are the terms:

The centre of mass is a point that acts as if all the mass was centered there (the mass on one side of the point is equal to the mass on the opposite side). If supported at the centre of mass, an object will be balanced under the influence of gravity.

The centre of gravity is the point at which where the sum (vector) of the gravitational forces act on an object which will be balanced on that point.

53. What are the differences between Weight and Mass?

Answer: Mass is a measure of how much matter an object has. Mass is specified in Gram or Kilograms.

Weight is a measure of how strongly gravity pulls on that matter. Force is measured in Newton's.

$$F = m.g$$

Thus if you were to travel to the moon your weight would change because the pull of gravity is weaker there than on Earth but, your mass would stay the same because you are still made up of the same amount of matter.

54. What is the difference between Projectile motion and a Rocket motion?

Answer: A projectile has no motor/rocket on it, so all of its momentum is given to it as it is launched. An example of a projectile would be pen that you throw across a room.

A rocket or missile does have a motor/rocket on it so it can accelerate itself while moving and so resist other forces such as gravity. In mechanics point of view projectile don't have any particular shape it is a point mass. Whereas rocket has a particular shape and hence it has centre of gravity situated at particular point on its body. Therefore rocket motion comes under kinetics and projectile comes under kinematics.

55. What type of cooling used in High Voltage Transformer?

Answer: The big transformers you find on power poles usually use oil as a dielectric insulator, in smaller HV transformers, like the ones in TV's are usually filled with resin.

56. What is honing of Cylinder Liners?

Answer: The honing equipment used has been manufactured by "Chris Marine". The head of the honing device consists of four synchronized stones. For the initial honing diamond stones are used to break up the hardened surface in the scuffed areas. For the main honing very coarse and hard stones are used to produce a very desirable rough surface all over the liner. The advantage, especially for the 2-stroke engines, is possibility to save the liner after a seizure, scuffing or blow-by or even to eliminate the ovality of the liner. Another advantage is that it is possible to machine a rough liner wall to obtain a well oiled surface.

57. What is the difference between Speed and Velocity?

Answer:

Speed is scalar quantity and **Velocity** is a vector.

Velocity has both speed and direction. Speed is expressed as distance moved (d) per unit of time (t). Speed is measured in the same physical units of measurement as velocity, but does not contain an element of direction. Speed is thus the magnitude component of velocity.

58. What is the difference between Yield and Ultimate tensile strength?

Answer: The yield strength is reached when the material becomes Non – linear (that is non elastic) and takes a permanent set when load is released. Material stretches but does not break. Ultimate strength is when it breaks and is higher than yield strength.

59. Difference between Yield Stress and Yield Strength?

Answer: Stress is a measure of the load applied to a sample relative to a cross sectional area of the sample. Strength is a quantification of the samples ability to carry a load.

The terms "yield strength" and "yield stress" of a material are usually used interchangeably (correct or not). It is the stress which will just cause the material to plastically deform. If a material yields at

30,000 psi, the yield stress is 30,000 psi. If the part in question has a cross sectional area of 2 square inches, the strength at yield would be 60,000 pounds, but usually we just say the yield strength is 30,000 psi.

60. What is the difference between Flexural strength and Tensile strength?

Answer: Flexural strength is resistance offered against bending. Tensile strength is resistance offered against tensile force.

61. What is the difference between Shear and Tensile strength?

Answer: Tensile Strength for a Bolt is determined by applying a Force along it long axis. Shear Strength for a Bolt is determined by applying a Force across its diameter, as it would be loaded in a lug joint. Tensile strength is strength in tension when pulling force is applied. And shear strength is strength against cutting force which is known as shear force.

62. What is the difference between Tensile strength and tensile modulus?

Answer: Tensile strength is the ultimate capacity of the material to resist a tensile load regardless of deflection.

Tensile modulus also known as Young's modulus is a measure of the stiffness of an isotropic elastic material. It is defined as the ratio of the uni-axial stress over the uni-axial strain. It is determined from the slope of a stress-strain curve traced during tensile tests conducted on a sample of the material.

63. What is OEE?

Answer: OEE means Overall Equipment Effectiveness.

This terminology widely used in Total productive maintenance, which is used to calculate the effectiveness of machines in manufacturing. Basically it captures the losses of machines in production and tries to improve defects on machines. Higher the OEE, more capable is the machine.

64. Why Involutes Curve used in Gear?

Answer: Involute curve is the path traced by a point on a line as the line rolls without slipping on the circumference of a circle. Involute curve has a contact angle between two gears when the tangents of two gears pass through the contact point without friction.

65. What is bearing stress?

Answer: The stress which acts on the contact surface area between two members is known as bearing stress. An example for this is the stress between nut and the washer

66. Which is hard material Cast Iron or Mild Steel?

Answer: Cast iron. Due to the excess carbon content than mild steel it is harder. The more carbon content, the more hardness will be. But it reduces the Weldability due to this hardness. It is brittle too.

67. What are the materials used for Sliding Wear pad?

Answer:
1. Urethane-Coated Base Material Provides Optimum Sliding Surfaces for Maximum Wear Ability.
2. Galvanized Steel for Backing on Head and Side Pads.
3. For crane services, we use Velcro Nylon wearing pad

68. How will you calculate the tonnage of Mechanical Press?

Answer:

$F = (S \times L \times T)/ 1000$

F = Force in kilo Newton's

S = shear stress of material in MPa

L = the total length of peripheries being cut/ sheared in mm

T = thickness of material in mm

69. What is the difference between a Shaper machine and a Planner machine?

Answer: In Shaper machine tool is having reciprocating motion and work piece is clamped on table which is stationary. It is mostly suitable for light duty operation. In Shaping large cutting force is transferred to tool. In Planer machine tool is having stationary and work piece is clamped on table which is reciprocating motion. It is mostly suitable for Heavy duty operation. In planner large cutting force is transferred to table.

70. What is the composition of Grey cast iron Grade 20?

Answer:
Carbon: 3.10 – 3.25%, Silicon: 1.75 – 1.95%, Manganese: 0.50 – 0.7%, Sulphur: 0.05 – 0.07%, Phosphorous: 0.04 – 0.07%

71. What is the composition of Cast iron Grade 35?

Answer: Carbon = 2.90 – 3.10%, Manganese = 0.60 – 1.00%, Silicon = 1.50 – 1.90%, Sulphur = 0.10%, Phosphorus = 0.15%, Chromium = 0.30%, Molybdenum = 0.30%, Cupper = 0.25%

72. What is the function of a thrust bearing?

Answer: Thrust bearings keep the rotor in its correct axial position.

73. What are the super alloys?

Answer: Super alloys are an alloy that exhibits excellent mechanical strength and creep resistance at high temperatures, having good surface finish.

74. What kinds of NDT methods are available?

Answer:
1. Visual Inspection
2. Microscopy inspection
3. Radiography Test
4. Dye Penetrate technique
5. Ultrasonic testing
6. Magnetic Particle inspection
7. Eddy Current technology
8. Acoustic Emission
9. Thermograph
10. Replica Metallographic

75. What is Stress Corrosion cracking?

Answer: Stress corrosion cracking (SCC) is a process involving the initiation of cracks and their propagation, possibly up to complete failure of a component, due to the combined action of tensile mechanical loading and a corrosive medium.

76. What is meant by D2 Material used for Die tooling?

Answer: D2 – High Carbon Cold Work Tool Steel.
D2 is a high Carbon, high Chromium, Molybdenum, Vanadium, Air hardening alloy tool steel which offers good wear resistance, high surface hardness, through hardening properties, dimensional stability and high resistance to tempering effect. D2 tool steel is also suitable for vacuum hardening.
Typical Composition:
C. = 1.50%, Si. = 0.30%, Cr. = 12.00%, Mo. = 0.80%, V. = 0.90%

77. What is Vacuum Induction Melting?

Answer: As the name suggests, the process involves melting of a metal under vacuum conditions. Electromagnetic induction is used as the energy source for melting the metal. Induction melting works by inducing electrical eddy currents in the metal. The source is the induction coil which carries an alternating current. The eddy currents heat and eventually melt the charge.

78. What is the difference between Bolt and Screw?

Answer: The main difference was based on the load acting on it, and the size. For smaller loads, screws are enough but in case of greater-loads, bolts are to be used. In bolt we give centrifugal force or tangential force and screw we give axial force for driving.

79. What is the significance of Torque (in N-m) given in the engine specification?

Answer: It give the moment about any point or simple rotation.

80. What are the uses of Graphite electrode in various fields?

Answer: Graphite electrode is used in EDM and in battery cells. It is also used in electric arc furnaces to melt the steel.

81. Why the Super alloys used for land based turbines?

Answer: Super alloys are the top most alloys used for their excellent strength and corrosion resistance as well as oxidation resistance. No other alloys can compete with these grades.

82. Difference between TIG & MIG welding

Tungsten inert gas welding-non consumable electrode
MIG-Metal inert gas welding-uses consumable electrode

83. Do you know epicyclic gear box? What is the practical application of epicyclic gear box?

Answer: Epicyclic gear box consists of sun gear planetary gears and an annular called ring gear. Different speed ratios are obtained by locking any one gear. If you lock any two gears, direct gear will be obtained. Mostly used in over drives.
Wrist watch is a practical application of epicyclic gear box

84. What is the purpose of scrapper ring?

Answer: Scrap the excess lube oil from the cylinder walls. There by preventing oil from entering combustion zone.

85. What is the difference between S.S to EN8?

Answer:

> SS- Stainless steel
> En- Medium carbon steel
> SS is Non Magnetic material & EN8 is Magnetic material
> SS is Corrosion resistant & EN8 is Magnetic material

86. How to calculate the speed of conveyer in Meter per Minute

Answer: Measure the diameter of the rollers around which the conveyor belt is wrapped. Multiply the diameter of the roller by pi ($\pi = 3.14159$). This calculation will yield the circumference of the rollers. Every time the roller spins one revolution, the conveyor will be moved a linear distance equivalent to the circumference of the roller. Pi is a dimensionless factor, meaning it does not matter whether inches, centimetres or any other units of measurement are used. Measure the revolutions per minute (RPM) of the rollers. Count how many full revolutions (rotations) are made by the roller in one minute. Multiply the RPM by the circumference of the roller. This calculation will give the linear distance traversed by a point on the conveyor belt in one minute.

87. What is the use of a PULLEY?

Answer: transmission of power (force) in rotary form

88. Why does cycle rim don't bend even in heavy loads?

Answer: Because of rubber tires. The load is distributed and its effect reduces i.e. tires absorbs heavy load and shocks with the support of steel rim. The rim has many spokes. The spokes distribute the load equally and the rubber tires absorb more than half of the load.

89. What is caustic embrittlement?

Answer: It is the actual physical change in metal that makes it extremely brittle and filled with minute cracks. It occurs particularly in the seams of riveted joints and around the rivet holes.

90. What type of section of same area will resist maximum bending moment?

Answer: I- section of same area resist more bending moment than a rectangular or circular section. The reason is obvious. In i-section larger area is concentrated at larger distance from neutral axis and hence stressed more.

In circular section, large area is concentrated near neutral axis and hence it is inefficient in resisting bending.

Design of Machine Elements: Basic

01. Define Ductility.

Answer: It is the property of the material enabling it to be drawn into wire, with the application of tensile force. It must be both strong and plastic. It is usually measured in terms of percentage elongation and reduction in area, (e.g.) Ni, Al, and Cu.

02. Define fatigue.

Answer: When a material is subjected to repeated stress, it fails at stresses below the yield point stress; such type of failure of the material is called fatigue.

03. Define: Factor of safety.

Answer: The ratio between maximum stresses to working stress is known as factor of safety. Factor of safety = Maximum stress /Working stress

04. Define endurance limit.

Answer: Endurance limit is the maximum value of completely reversed stress that the standard specimen can sustain an infinite number (106) of cycles without failure.

05. What is impact load?

Answer: If the time of load application is less than one third of the lowest natural period of vibration of the part, it is called an impact load.

06. What are the various phases of design process?

Answer: The various phases of design process are:
1. Recognition of need.
2. Definition of problem
3. Synthesis
4. Analysis and optimization
5. Evaluation
6. Presentation

07. What is the use of Goodman & Soderberg diagrams?

Answer: They are used to solve the problems of variable stresses.

08. What are the factors affecting endurance strength.

Answer: Factors affecting endurance strength are
1. Load
2. Surface finish
3. Size
4. Temperature
5. Impact
6. Reliability

09. What are the types of variable stresses?

Answer: types of variable stresses are:
1. Completely reversed or cyclic stresses
2. Fluctuating stresses
3. Repeated stresses

10. Differentiate between repeated stress and reversed stress.

Answer: Repeated stress refers to a stress varying from zero to a maximum value of same nature.
Reversed stress of cyclic stress varies from one value of tension to the same value of compression.

11. What are the types of fracture?

Answer: The two types of fracture are
1. Ductile fracture
2. Brittle fracture

12. Distinguish between brittle fracture and ductile fracture.

Answer: In brittle fracture, crack growth is up to a small depth of the material.
In ductile fracture large amount of plastic deformation is present to a higher depth.

13. Define stress concentration and stress concentration factor.

Answer: Stress concentration is the increase in local stresses at points of rapid change in cross section or discontinuities.

Stress concentration factor is the ratio of maximum stress at critical section to the nominal stress

14. Explain size factor in endurance strength.

Answer: Size factor is used to consider the effect of the size on endurance strength. A large size object will have more defects compared to a small one. So, endurance strength is reduced. If K is the size factor, then

Actual endurance strength = Theoretical endurance limit × K

15. Explain Griffith theory. (Or) State the condition for crack growth.

Answer: A crack can propagate if the energy release rate of crack is greater than crack resistance.

16. What are the modes of fracture?

Answer: The different the modes of fractures are:
1. Mode I (Opening mode) – Displacement is normal to crack surface.
2. Mode II (Sliding mode) – Displacement is in the plane of the plate.
3. Mode III (Tearing mode) – Out of plane shear.

17. What are the factors to be considered in the selection of materials for a machine element?

Answer: While selecting a material for a machine element, the following factors are to be considered
1. Required material properties
2. Manufacturing ease
3. Material availability
4. Cost

18. What are various theories of failure?

Answer: The failure theories are:
1. Maximum principal stress theory.
2. Maximum shear stress theory.
3. Maximum principal strain theory.

19. List out the factors involved in arriving at factor of safety.

Answer: The factors involved in arriving at factor of safety are:
1. Material properties
2. Nature of loads
3. Presence of localized stresses

4. Mode of failures

20. Give some methods of reducing stress concentration.

Answer: Some of the methods are:
1. Avoiding sharp corners.
2. Providing fillets.
3. Use of multiple holes instead of single hole
4. Undercutting the shoulder parts.

21. Explain notch sensitivity. State the relation between stress concentration factor and notch sensitivity.

Answer: Notch sensitivity (q) is the degree to which the theoretical effect of stress concentration is actually reached.

$$K_f = 1 + q \, (K_t - 1)$$

22. What are the factors that affect notch sensitivity?

Answer: The factors effecting notch sensitivity are:
1. Material
2. Notch radius
3. Size of component
4. Type of loading
5. Grain Structure

23. What are the different types of loads that can act on machine components?

Answer: different loads on machine components are:
1. Steady load.
2. Variable load.
3. Shock load.
4. Impact load.

24. Define machinability

Answer: It is the property of the material, which refers to a relative ease with which a material can be cut. It is measured in a number of ways such as comparing the tool life for cutting different material

25. What is an S-N Curve?

Answer: An S- N curve has fatigue stress on 'Y' axis and number of loading cycles in 'X' axis. It is used to find the fatigue stress value corresponding to a given number of cycles.

26. What is curved beam?

Answer: In curved beam the neutral axis does not coincide with the centroidal axis.

27. Give some example for curved beam.

Answer: C frame, crane hook

28. What is principle stress and principle plane?

A plane which has no shear stress is called principle plane the corresponding stress is called principle stress.

29. Write the bending equation.

Answer: The bending moment equation is, $M/I = f/y = E/R$,
Where,

   M – Bending moment (M is in N-mm)
   I - Moment of inertia about centroidal axis (I is in mm^4)
   f – Bending Stress (f is in N/mm^2)
   y - Distance from neutral axis (y is in mm)
   E - Young's modulus (E is in N/mm^2)
   R - Radius of curvature (R is in mm)

30. Write the torsion equation.

Answer: The torsional equation is, $T/J = q/r = G\Theta/L$
Where,

   T – Torsional moment (T is in N-mm)
   J - Polar moment of inertia (J is in mm^4)
   q – Shear stress in the element (q is in N/mm^2)
   r - Distance of element from centre of shaft (r is in mm)
   G- Modulus of Rigidity (G is in N/mm^2)
   Θ – Angle of twist (Θ is in radians)
   L – Length of the shaft (L is in mm)

Design of Machine Elements: Springs

01. What is a spring?

Answer: A spring is an elastic member, which deflects, or distorts under the action of load and regains its original shape after the load is removed.

02. State any two functions of springs.

Answer:

   1. To measure forces in spring balance, meters and engine indicators.
   2. To store energy.

03. What are the various types of springs?
Answer:
1. Helical springs
2. Spiral springs
3. Leaf springs
4. Disc spring or Belleville springs

04. What are the requirements of spring while designing?
Answer:
1. Spring must carry the service load without the stress exceeding the safe value.
2. The spring rate must be satisfactory for the given application.

05. Classify the helical springs.
Answer:
1. Close – coiled or tension helical spring.
2. Open –coiled or compression helical spring.

06. Define Leaf springs
Answer: A leaf spring consists of flat bars of varying lengths clamped together and supported at both ends, thus acting as a simply supported beam.

07. What is spring index (C)?
Answer: The ratio of mean or pitch diameter to the diameter of wire for the spring is called the spring index.

08. Define Belleville Springs
Answer: They are made in the form of a cone disc to carry a high compressive force. In order to improve their load carrying capacity, they may be stacked up together. The major stresses are tensile and compressive.

09. What is pitch?
Answer: The axial distance between adjacent coils in uncompressed state.

10. What is solid length?
Answer: The length of a spring under the maximum compression is called its solid length. It is the product of total number of coils and the diameter of wire.

$$L_s = n_t \times d$$

Where, n_t = total number of coils.

11. What are the end conditions of spring?
Answer:
1. Plain end.

2. Plain and ground end
3. Squared end
4. Squared and ground end.

12. What is buckling of springs?
Answer: The helical compression spring behaves like a column and buckles at a comparative small load when the length of the spring is more than 4 times the mean coil diameter.

13. What is surge in springs?
Answer: The material is subjected to higher stresses, which may cause early fatigue failure. This effect is called as spring surge.

14. What is a laminated leaf spring?
Answer: In order to increase, the load carrying capacity, number of flat plates are placed and below the other.

15. What are semi elliptical leaf springs?
Answer: The spring consists of number of leaves, which are held together by U-clips.
The long leaf fastened to the supported is called master leaf. Remaining leaves are called as graduated leaves.

16. What is nipping of laminated leaf spring?
Answer: Pre-stressing of leaf springs is obtained by a difference of radii of curvature known as nipping.

17. What are the various applications of springs?
Answer: The springs are used in various applications, they are
    1. Used to absorb energy or shocks (e.g. shock absorbers, buffers, etc.)
    2. To apply forces as in brakes clutches, spring-loaded valves, etc.
    3. To measure forces as in spring balances and engine indicators
    4. To store energy as in watches

18. Define free length.
Answer: Free length of the spring is the length of the spring when it is free or unloaded condition. It is equal to the solid length plus the maximum deflection or compression plus clash allowance.

$$L_f = \text{solid length} + Y_{max} + 0.15\ Y_{max}$$

19. Define spring index.
Answer: Spring index (C) is defined as the ratio of the mean diameter of the coil to the diameter of the wire.

$$C = D/d$$

20. Define spring rate (stiffness).

Answer: The spring stiffness or spring constant is defined as the load required per unit deflection of the spring.

$$K = W/y$$

Where, W - Load

y - Deflection

21. Define pitch.

Answer: Pitch of the spring is defined as the axial distance between the adjacent coils in uncompressed state. Mathematically Pitch=free length/n-1

22. What are the points to be taken into consideration while selecting the pitch of the spring?

Answer: The points taken into consideration of selecting the pitch of the springs are

1. The pitch of the coil should be such that if the spring is accidentally compressed the stress does not increase the yield point stress in torsion.
2. The spring should not be close up before the maximum service load is reached.

23. Define active turns.

Answer: Active turns of the spring are defined as the number of turns, which impart spring action while loaded. As load increases the no of active coils decreases.

24. Define inactive turns.

Answer: Inactive turns of the spring are defined as the number of turns which does not contribute to the spring action while loaded. As load increases number of inactive coils increases from 0.5 to 1 turn.

25. What are the different kinds of end connections for compression helical springs?

Answer: The different kinds of end connection for compression helical springs are

1. Plain ends
2. Ground ends
3. Squared ends
4. Ground & square ends

26. Write about the eccentric loading of springs?

Answer: If the load acting on the spring does not coincide with the axis of the spring, then spring is said to be have eccentric load. In eccentric loading the safe load of the spring decreases and the stiffness of the spring is also affected.

27. Explain about surge in springs?

Answer: When one end of the spring is resting on a rigid support and the other end is loaded suddenly, all the coils of spring does not deflect equally, because some time is required for the propagation of stress along the wire. Thus a wave of compression propagates to the fixed end from where it is reflected back to the deflected end this wave passes through the spring indefinitely. If the time interval between the load application and that of the wave to propagate are equal, then resonance will occur. This will result in very high stresses and cause failure. This phenomenon is called surge.

28. What are the methods used for eliminating surge in springs?
Answer: The methods used for eliminating surge are
1. By using dampers on the centre coil so that the wave propagation dies out
2. By using springs having high natural frequency.

29. What are the disadvantages of using helical spring of non-circular wires?
Answer:
1. The quality of the spring is not good
2. The shape of the wire does not remain constant while forming the helix. It reduces the energy absorbing capacity of the spring.
3. The stress distribution is not favourable as in circular wires. But this effect is negligible where loading is of static nature.

30. Why concentric springs are used?
Answer:
1. To get greater spring force within a given space
2. To insure the operation of a mechanism in the event of failure of one of the spring

31. What is the advantage of leaf spring over helical spring?
Answer: The advantage of leaf spring over helical spring is that the end of the spring may be guided along a definite path as it deflects to act a structural member in addition to energy absorbing device.

32. Write notes on the master leaf & graduated leaf?
Answer: The longest leaf of the spring is known as main leaf or master leaf has its ends in the form of an eye through which bolts are passed to secure the spring. The leaf of the spring other than master leaf is called the graduated leaves.

33. What is meant by nip in leaf springs?
Answer: By giving greater radius of curvature to the full length leaves than the graduated leaves, before the leaves are assembled to form a spring thus a gap or clearance will be left between the leaves. This initial gap is called nip.

34. What is the application of leaf spring?
Answer: The leaf springs are used in automobiles as shock absorbers for giving suspension to the automobile and it gives support to the structure.

35. Define flat spiral spring.
Answer: A flat spiral spring is a long thin strip of elastic material wound like a spiral. These springs are frequently used in watch springs, gramophones, etc

36. What are the differences between helical torsion spring and tension helical springs?
Answer: Helical torsion springs are wound similar to that of tension springs but the ends are shaped to transmit torque. The primary stress in helical torsion spring is bending stress whereas in tension springs the stresses are torsional shear stresses.

37. Define helical springs.
Answer: The helical springs are made up of a wire coiled in the form of a helix and are primarily intended for compressive or tensile load.

38. What are the different types of helical springs?
Answer: The different types of helical springs are
1. Open coil helical spring
2. Closed coil helical spring

39. What is closed coil helical spring?
Answer:
1. The spring wires are coiled very closely, each turn is nearly at right angles to the axis of helix
2. Helix angle is small, i.e. less than 10 Degree.

40. What is open coil helical spring?
Answer:
1. The wires are coiled such that there is a gap between the two consecutive turns.
2. Helix angle is large i.e. greater than 10 degree.

Design of Machine Elements:
Shafts, Keys, Bolts, Nuts, Screws, Fasteners, Bearings, Flywheel & Governor

01. Define the term critical speed.
Answer: The speed, at which the shaft runs so that the additional deflection of the shaft from the axis of rotation becomes infinite, is known as critical or whirling speed.

02. Factors considered designing a shaft?
Answer: Strength and Stiffness

03. How does the function of flywheel differ from that of governor?

Answer: A governor regulates the mean speed of an engine when there are variations in the mean loads. It automatically controls the supply of working fluid to engine with the varying load condition & keeps the mean speed with contain limits. It does not control the speed variation caused by the varying load. A flywheel does not maintain const speed.

04. What is a key? What are the types of keys?

Answer: A key is device, which is used for connecting two machine parts for preventing relative motion of rotation with respect to each other.

The different types of keys are

1. Saddle key
2. Tangent key
3. Sunk key
4. Round key and taper pins

05. What is the main use of woodruff keys?

Answer: A woodruff key is used to transmit small value of torque in automotive and machine tool industries. The keyway in the shaft is milled in a curved shape whereas the key way in the hub is usually straight.

06. What are the various failures occurred in sunk keys?

Answer: Shear failure and Crushing failure.

07. What is the function of a coupling between two shafts?

Answer: Couplings are used to connect sections of long transmission shafts and to connect the shaft of a driving machine to the shaft of a driven machine.

08. Under what circumstances flexible couplings are used?

Answer: They are used to join the abutting ends of shafts when they are not in exact alignment. They are used to permit an axial misalignment of the shafts without under absorption of the power, which the shafts are transmitting.

09. What are the purposes in machinery for which couplings are used?

Answer: Couplings are used

1. To provide the connection of shafts of units those are manufactured separately such as motor and generator and to provide for disconnection for repairs or alterations.
2. To provide misalignment of the shafts or to introduce mechanical flexibility.
3. To reduce the transmission of shock from one shaft to another.
4. To introduce protection against over load.

10. What are the main functions of the knuckle joints?

Answer: It is used to transmit axial load from one machine element to other.

11. How is a bolt designated?

Answer: A bolt is designated by a letter M followed by nominal diameter and pitch in mm.

12. What factors influence the amount of initial tension?

Answer: The factors influence the amounts of initial tension are
1. External load
2. Material used
3. Bolt diameter

13. What is bolt of uniform strength?

Answer: A bolt of uniform strength has equal strength at the thread and shank portion.

14. What are the ways to produce bolts of uniform strength?

Answer: the ways to produce bolts of uniform strength are
1. Reducing shank diameter equal to root diameter.
2. Drilling axial hole

15. What stresses act on screw fastenings?

Answer: stresses act on screw fastenings are
1. Initial stresses due to screwing up
2. Stresses due to external forces
3. Combined stresses.

16. What are the different applications of screwed fasteners?

Answer: The different applications of screwed fasteners are
1. For readily connecting & disconnecting machine parts without damage
2. The parts can be rigidly connected
3. Used for transmitting power

17. What are the advantages of screwed fasteners?

Answer: The advantages of screwed fasteners are
1. They are highly reliable in operation
2. They are convenient to assemble & disassemble
3. A wide range of screws can be used for various operating conditions
4. They are relatively cheap to produce.

18. Define pitch.

Answer: Pitch is defined as the distance from appoint on one thread to the corresponding on the adjacent thread in the same axis plane.

19. Define lead.

Answer: Lead is defined as the distance, which a screw thread advances axially in one rotation of the nut.

20. What are the types of welded joints?

Answer: the types of welded joints are

1. Butt joint
2. Lap joint
3. T – joint
4. Corner joint
5. Edge joint.

21. What are the different types of metric thread?

Answer: the different types of metric thread are

1. BSW (British standard Whit worth)
2. BSE (British standard End

22. Define welding.

Answer: Welding can be defined as a process of joining two similar or dissimilar metals with or without application of pressure along with or without addition of filler material.

23. What are the two types of stresses are induced in eccentric loading of loaded joint?

Answer: the two types of stresses are induced in eccentric loading of loaded joint are

1. Direct shear stress.
2. Bending or torsional shear stress.

24. Define butt and lap joint.

Answer:

Butt joint: The joint is made by welding the ends or edges of two plates.

Lap joint: The two plates are overlapping each other for a certain distance and then welded. Such welding is called fillet weld.

25. When will the edge preparation need?

Answer: If the two plates to be welded have more than 6mm thickness, the edge preparation should be carried out.

26. What are the two types of fillet weld?

Answer: Two types of fillet weld are

1. Longitudinal or parallel fillet weld
2. Transverse fillet weld

27. State the two types of eccentric welded connections.

Answer: two types of eccentric welded connections are
   1. Welded connections subjected to moment in a plane of the weld.
   2. Welded connections subjected to moment in a plane normal to the plane of the weld.

28. What are the practical applications of welded joints?

Answer: It has employed in manufacturing of machine frames, automobile bodies, aircraft, and structural works.

29. What is bearing?

Answer: Bearing is a stationery machine element which supports a rotating shafts or axles and confines its motion.

30. Classify the types of bearings.

Answer: the types of bearings are

(A) Depending upon the type of load coming upon the shaft:
   1. Radial bearing
   2. Thrust bearings.

(B) Depending upon the nature of contact:
   1. Sliding contact
   2. Rolling contact bearings or Antifriction bearings.

31. What are the required properties of bearing materials?

Answer: Bearing material should have the following properties.
   1. High compressive strength
   2. Low coefficient of friction
   3. High thermal conductivity
   4. High resistance to corrosion
   5. Sufficient fatigue strength
   6. It should be soft with a low modulus of elasticity
   7. Bearing materials should not get weld easily to the journal material.

32. What is a journal bearing?

Answer: A journal bearing is a sliding contact bearing which gives lateral support to the rotating shaft.

33. What is known as self – acting bearing?

Answer: The pressure is created within the system due to rotation of the shaft; this type of bearing is known as self – acting bearing.

34. What are the types of journal bearings depending upon the nature of contact?

Answer: The types of journal bearings are

1. Full journal bearing
2. Partial bearing
3. Fitted bearing.

35. What are the types of journal bearing depending upon the nature of lubrication?
Answer: The types are
1. Thick film type
2. Thin film type
3. Hydrostatic bearings
4. Hydrodynamic bearing.

36. What is flywheel?
Flywheel is a machine elements used to minimize the fluctuation of speed in a engine.

37. What is the function of flywheel?
Answer: A flywheel used in machine serves as a reservoir which stores energy during the period when the supply of energy is more than the requirement and releases it dulling the period when the requirement of energy is more than the supply.

38. Define the term 'fluctuation of speed' and 'fluctuation of energy'.
Answer: The ratio of maximum fluctuation of speed to the mean speed is called co efficient of fluctuation of speed. The ratio of fluctuation of energy to the mean energy is called coefficient of fluctuation of energy.

39. State the type of stresses induced in a rim flywheel?
Answer: the stresses induced in a rim flywheel are
1. Tensile stress due to centrifugal force
2. Tensile bending stress caused by the restraint of the arms and
3. The shrinkage stresses due to unequal rate of cooling of casting.

40. What are the stresses induced in flywheel arms?
Answer: The stresses induced in flywheel arms are
1. Tensile stress due to centrifugal force.
2. Bending stress due to torque.
3. Stress due to belt tension.

Fluid Mechanics & Fluid Machinery:
Turbine, Boiler, compressor, Pumps, Bearings Etc...

01. What is the difference between streamline body and bluff body?
Answer: In streamline body the shape is such that separation in flow occurs past the near most part of the body so that wake formed is small and thus friction drag is much greater than pressure drag. In bluff body the flow gets separated much ahead of its rear resulting in large wake and thus pressure drag is much greater than the friction drag.

02. Define hydraulically efficient channel cross section.
Answer: The shape of such section is that which produces minimum wetted perimeter for a given area of flow and carries maximum flow.

03. What is the difference between Center of Mass and Centroid?
Answer: If the material composing a body is uniform or homogeneous, the density or specific weight will be constant throughout the body, and then the Centroid is the same as the centre of gravity or centre of mass.
Centroid: Centroid is the point, where the whole area of plane is going to be act. It is valid only for 2D problems like plane figures, square plate etc. The centre of mass is a point that acts as if all the mass was centered there (the mass on one side of the point is equal to the mass on the OPPOSITE side). If supported at the centre of mass, an object will be balanced under the influence of gravity.

04. On what factors does the pressure at a point as a static mass of liquid depends upon?
Answer: Specific weight of liquid and the depth below the free liquid surface.

05. When maximum discharge is obtained in nozzle?
Answer: At the critical pressure ratio.

06. Under what condition the work done in reciprocating compressor will be least?
Answer: It is least when compression process approaches isothermal. For this purpose, attempts are made to cool the air during compression.

07. What is the difference between stalling and surging in rotary compressions?
Answer: Stalling is a local phenomenon and it occurs when How breaks away from the blades. Surging causes complete breakdown of flow and as such it affects the whole machine.

08. State Archimedes principle.
Answer: Any weight, floating or immersed in a liquid, is acted upon by a buoyant force equal to the weight of the liquid displaced. This force acts through the centre of buoyancy, i.e. the e.g. of the displaced liquid.

09. What do you understand by centre of buoyancy?

Answer: Centre of buoyancy is the centre of gravity of the displaced liquid and buoyant force acts through it.

10. Why the Centrifugal Pump is called High Discharge pump?

Answer: Centrifugal pump is a kinetic device. The centrifugal pump uses the centrifugal force to push out the fluid. So the liquid entering the pump receives kinetic energy from the rotating impeller. The centrifugal action of the impeller accelerates the liquid to a high velocity, transferring mechanical (rotational) energy to the liquid. So it discharges the liquid in high rate. It is given in the following formulae:

Centrifugal force $F = (MV^2)/R$.

Where, M-Mass, V-Velocity, R-Radius

11. Why the electric motor of a fan with backward curved blades is never got overloaded under any condition?

Answer: The maximum power is consumed at about 70% of maximum flow in case of fan with backward blades. For higher flow, power consumption gets lower.

12. Why the work per kg of air flow in axial flow compressor is less compared to centrifugal compressor for same pressure ratio?

Answer: Isentropic efficiency of axial flow compressor is higher.

13. What is pitting? How it is caused?

Answer: Non uniform corrosion over the entire metal surface, but occurring only in small pits is called pitting. It is caused by lack of uniformity in metal.

14. How Cavitation can be eliminated by Pump?

Answer: Cavitation means bubbles are forming in the liquid. To avoid Cavitation, we have to increase the Pump size to one or two inch to increase the pressure of the Suction Head, or decrease the Pump Speed.

15. Which impurities form hard scale and which impurities soft scale?

Answer: Sulphates and chlorides of lime and magnesium form hard scale, and carbonates of lime and magnesium form soft scale.

16. What is the difference between hard water and soft water?

Answer: Hard water contains excess of scale forming impurities and soft water contains very little or no scale forming substances.

17. Which two elements in feed water can cause corrosion of tubes and plates in boiler?

Answer: Acid and oxygen in feed water lead to corrosion.

18. Why Cavitation will occur in Centrifugal Pump and not in Displacement Pump?

Answer: The formation of cavities (or bubbles) is induced by flow separation, or non-uniform flow velocities, inside a pump casing. In centrifugal pumps the eye of the pump impeller is smaller than the flow area of pipe. This decrease in flow area of pump results in increase in flow rate. So pressure drop happened between pump suction and the vanes of the impeller. Here air bubbles or cavities are formed because of liquid vapour due to increase in temperature in impeller. This air bubbles are transmitted to pump which forms Cavitation.

19. Why large boilers are water tube type?
Answer: Water tube boilers raise steam fast because of large heat transfer area and positive water circulation. Thus they respond faster to fluctuations in demand. Further single tube failure does not lead to catastrophy.

20. What type of boiler does not need a steam drum?
Answer: Supercritical pressure boiler.

21. Why manholes in vessels are usually elliptical in shape?
Answer: Elliptical shape has minimum area of opening and thus plate is weakened the least. Further it is very convenient to insert and take out the cover plate from elliptical opening.

22. Which Pump is more Efficient Centrifugal Pump or Reciprocating Pump?
Answer: Centrifugal pump. Because of flow rate is higher compared to reciprocating pump. Flow is smooth and it requires less space to install. Lower initial cost and lower maintenance cost.

23. Why Centrifugal Pump is not called as a Positive Displacement Type of Pump?
Answer: The centrifugal has varying flow depending on pressure or head, whereas the Positive Displacement pump has more or less constant flow regardless of pressure.
Likewise viscosity is constant for positive displacement pump where centrifugal pump have up and down value because the higher viscosity liquids fill the clearances of the pump causing a higher volumetric efficiency. When there is a viscosity change in supply there is also greater loss in the system. This means change in pump flow affected by the pressure change.
One more example is, positive displacement pump has more or less constant efficiency, where centrifugal pump has varying efficiency rate.

24. Low water in boiler drum is unsafe because it may result in overheating of water tubes in furnace. Why it is unsafe to have high water condition in boiler drum?
Answer: High drum level does not allow steam separation to be effective and some water can be carried over with steam which is not desirable for steam turbine.

25. What is the difference between Critical Speed and Whirling Speed?
Answer: In Solid mechanics, in the field of rotor dynamics, the critical speed is the theoretical angular velocity which excites the natural frequency of a rotating object, such as a shaft, propeller or gear. As the speed of rotation approaches the objects natural frequency, the object begins to resonate which dramatically increases system vibration. The resulting resonance

occurs regardless of orientation. Whirling Speed is due to the unbalanced forces acting on a rotating shaft.

26. Maximum combustion temperature in gas turbines is of the order of 1100 to 10°C whereas same is around 0°C in I.C. engine? Why?
Answer: High temperature in I.C. engine can be tolerated because it lasts for a fraction of second but gas turbines have to face it continuously which metals can't withstand.

27. What is basic difference between impulse turbine and reaction turbine?
Answer:
- (A) In impulse turbine, jet is used to create impulse on blades which rotates the turbine and in reaction turbine, no jet is used pressure energy is converted into kinetic energy.
- (B) In impulse turbine fluid enter& leave with same energy, but in reaction turbine fluid enter with pressure energy& leaves with kinetic energy
- (C) (C)In impulse turbine all the pressure drops in nozzle only & in reaction turbine pressure drops both fixed & moving blades. The difference is due to blade profiles.

28. What are the causes of failure of superheater tubes?
Answer: Superheater tubes are subjected to the most severe combination of stress, temperature and corrosive environment. In addition to high temperature strength, resistance to corrosion is also important. For example, low alloy ferrite steel such as 1% Cr, 1% Mo would not be used at metal temperatures above 580°C because of inadequate resistance to corrosion and oxidation over a full service life of 100,000/150,000 hr.
Failures in superheater tubes may arise from:
- (A) Prior fabrication history
- (B) Faulty heat treatment
- (C) Consequences of welding
- (D) Overheating of the tube metal
- (E) Gas-side corrosion
- (F) Stress corrosion (austenitic steels).

29. Out of electric heater and heat pump, which is economical in operation?
Answer: Heat pump.

30. What is the Difference between a Generator and Inverter?
Answer: An inverter is only effective if there is already a source of electrical energy. It cannot generate its own. It can simply convert electrical energy that is already there. On the other hand, a traditional generator cannot make AC current into DC current.

31. Why is sound faster in warm air?
Answer: The speed of sound in air $C_{air} = 331.3 + (0.66 \times T)$ m/s, where T is the temperature in °C.

The speed of sound is proportional to gas temperature and inversely proportional to its molar mass.

Sound is transferred by collisions of molecules. Therefore sound waves will travel faster on warm air because collision of molecules of air in warm air is greater.

32. What is a Newtonian fluid?

Answer: A Newtonian fluid possesses a linear stress strain relationship curve and it passes through the origin. The fluid properties of a Newtonian fluid do not change when any force acts upon it.

33. What are the causes of failure of boiler tubes?

Answer: Boiler tubes, usually are made from carbon steel and are subject to
  - (A) High rates of heat transfer
  - (B) (B). bending stresses due to uneven heating, especially at expanded or welded joints into headers or drums,
  - (C) External erosion from burners and flue gas,
  - (D) Possible corrosion on the boiler side, and
  - (E) Occasional manufacturing defects.

Failure may occur due to following reasons:
  - (A) High thermal ratings may lead to rapid failure if the internal fluid flow is reduced for any reason. The resultant overheating leads to a failure by creep, characterised by the bulging of the tube with the eventual development of a longitudinal split.
  - (B) Fatigue cracking due to bending stresses occurs. These are associated with change of section and/or weld undercut, where tubes are expanded or welded into headers.
  - (C) Failure may arise due to overstressing of a reduced section of metal.
  - (D) Sudden failure of the boiler tube due to corrosion arises from embrittlement of the carbon steel due to interaction between atomic hydrogen from the corrosion process and the iron carbide present in the steel.
  - (E) Defects in tube manufacture, although far from being a regular occurrence, can be a cause of serious trouble. Lamination in boiler tubes or score marks arising from the cold drawing of tubes give rise to premature failure and may promote corrosion at these regions.

34. Why supercritical boilers use less amount of steel compared to non supercritical boilers?

Answer: Supercritical boilers do not head heavy drum for separation of steam from mixture of water and steam.

35. What does a pump develops? Flow or Pressure?

Answer: A pump does not create pressure, it only creates flow. Pressure is a measurement of the resistance to flow.

36. What is a Hydrostatic system?

Answer:

Hydrostatics is the study of fluid bodies that are
  - (A) At rest

(B) Moving sufficiently slowly so there is no relative motion between adjacent parts of the body

For hydrostatic situations
   (A) There are no shear stresses.
   (B) There are only pressure forces that act perpendicular to any surface.
It's a closed loop hydraulic systems. It comprises of motor and pump. Here pump supplies energy to motor and motor gives return energy to pump supply.

37. What is the difference between Blower and Fan?
Answer: Fan is an air pushing device. Either Axial or Centrifugal type systems are used to move the air in low pressure. It is rotated by a motor separately.
When the fan is a housing of blades and motor, then it called as Blower. It directs the air in a single path with high pressure.

38. What is Hydrodynamic Cavitation?
Answer: Hydrodynamic Cavitation describes the process of vaporization in a constrained channel at a specific velocity. Bubble generation and Bubble implosion which occurs in a flowing liquid as a result of a decrease and subsequent increase in pressure.

39. Is gate valve used for Throttling?
Answer: Gate valves are not suitable for throttling because the control of flow is difficult for the valve's design, and the flow of fluid slapping against a partially open gate can cause extensive damage to the valve.

40. Where Multi-stage pump used
Answer: Pressure washing of Aircraft, Trains, Boats and Road vehicles as well as Spray washing of industrial parts and Electronic components.

41. Which furnace burns low ash fusion coal and retains most of the coal ash in the slag?
Answer: Cyclone furnace.

42. What is the effect of friction on flow of steam through a nozzle?
Answer: To decrease both mass flow rate and wetness of steam.

43. How the thickness of thermal boundary layer and thickness of hydrodynamic boundary layer related?
Answer: Ratio of their thickness = (Prandtl number)-1/3.

44. What is the function of Scoop in BFP (Boiler Feed Water Pump) in Thermal Power Station?
Answer: The Function of Scoop tube is regulating the varying amount of oil level in the coupling during operation of infinite variable speed.

45. In the Thermal Power Plant why Deaerator (D/A) placed on height?

Answer: To build a Very high pressure and the temperature for a boiler feed water pump and it discharge high pressure water to the boiler. And to provide the required Net Positive Suction Head (NPSH) for the BFW pump and to serve as a storage tank to ensure a continuous supply of feed water during rapid changes in BFP.

46. In orifice why the Pressure and Temperature are decreases?

Answer: Orifice is a small hole like a nozzle. When a high pressure fluid passes through the orifice,
Pressure gets reduced suddenly and the velocity of the fluid gets increased. Also the heat transfer rate increases. We know that Heat transfer rate is directly proportional to the difference in temperature, Area and the Heat transfer coefficient. Heat transfer Coefficient remains constant for a fluid at a particular temperature.
$$Q= KA \ (T_1-T_2)$$
If the heat transfer rate increases, it seems the difference in temperature gets increased. There is no way in reduction of inlet temperature of the orifice. As a result, the outlet temperature of the orifice gets reduced. Hence the pressure and temperature gets reduced when it passes through orifice

47. Why gas turbine power plant needs efficient compressor?

Answer: Because a large portion of turbine work is eaten away by compressor and its inefficiency will affect net power output and cost of generation.

48. Why rockets using liquid hydrogen have higher specific impulse compared to liquid hydrocarbon?

Answer: Liquid hydrogen has higher burning velocity.

49. Why axial flow compressor is preferred for gas turbines for aeroplanes?

Answer: Because it has low frontal area.

50. What is the difference between gas turbine and a steam turbine?

Answer: Gas turbine works on Brayton cycle where as steam turbine works Rankine cycle. Construction, operation of a gas turbine is entirely different to steam turbine. Gas turbine has a compressor to compress the combustion air, a combustion chamber to burn the fuel and a turbine section to extract the work for burning fuel. Steam turbine is just has a turbine section to extract the work from steam.

51. What is operating pressure?

Answer: The amount of pressure nearest the point of performing work at the output end of a pneumatic system. The system operating pressure is used to specify the capability of valves and actuators.

52. What is the effect of inter cooling in gas turbines?

Answer: It decreases thermal efficiency but increases net output.

53. What are the safety valves? How many on each boiler?

Answer: A valve opening automatically to relieve excessive pressure, especially in a boiler. There are normally two to six safety valves provided in the drum depending upon the capacity. The super heater outlet will have one to three safety valves on either side of the boiler. There will be an electrometric relief valve on the super heater pipe in addition. This valve will be set at lower pressure than the lowest set safety valve on the super heater. The reheater pipes both at the inlet and outlet side will also have safety valves which can range from two to eight both in the inlet and outlet of the reheater put together.

54. What is a sentinel valve?

Answer: Sentinel valves are simply small relief valves installed in some systems to warn of impending over pressurization. Sentinel valves do not relieve the pressure of the system. If the situation causing the sentinel valve to lift is not corrected, a relief valve (if installed) will lift to protect the system or component. If a relief valve is not installed, action must be taken quickly to secure the piece of equipment or system to reduce the pressure.

55. In which reactor the coolant and moderator are the same?
Answer: Pressurised water reactor.

56. Which reactor has no moderator?
Answer: Fast breeder reactor.

57. What are thermal neutrons?
Answer: Thermal neutrons are slow neutrons (having energy below 1 eV) which are in thermal equilibrium with their surroundings.

58. What is the function of Hydrogen seals work on a generator?

Answer: Provide a seal between the generator housings and rotor shaft to maintain the pressurized hydrogen gas inside the generator. Also, provides a trap-vent system to prevent the release of hydrogen into the turbine generator lube oil system and building atmosphere.

59. How is the excess discharge pressure prevented?

Answer: Pressure relief valves on the discharge side of each seal oil pump relieve back to seal oil system.

60. Which two seal oil pumps are driven from the same motor?

Answer: Main seal oil pump and re-circular seal oil pump; both are driven by Main seal oil pump motor.

61. Which heating value is indicated by a calorimeter and why?
Answer: Gross heating value because steam is condensed and heat of vapour formed is recovered.

62. When does Emergency Seal Oil Pump automatically start?
Answer: When its pressure switch senses Main Seal Oil Pump discharge pressure reduced to 78 PSI.

63. What is the consequence of not maintaining hydrogen (or air) pressure in generator casing at a value above atmospheric pressure when seal oil system is in service?

Answer: Failure to do so will cause excessive seal oil to be drawn into the generator.

64. What is a radial-flow turbine?

Answer: In a radial-flow turbine, steam flows outward from the shaft to the casing. The unit is usually a reaction unit, having both fixed and moving blades.

65. What are four types of turbine seals?

Answer:
1. Carbon rings fitted in segments around the shaft and held together by garter or retainer springs.
2. Labyrinth mated with shaft serration's or shaft seal strips.
3. Water seals where a shaft runner acts as a pump to create a ring of water around the shaft. Use   only treated water to avoid shaft pitting.
4. Stuffing box using woven or soft packing rings that are compressed with a gland to prevent leakage along the shaft.

66. What are two types of clearance in a turbine?

Answer:
Radial – clearance at the tips of the rotor and casing.
Axial – the fore-and-aft clearance, at the sides of the rotor and the casing.

67. Which reactor uses natural uranium as fuel?

Answer: Gas cooled reactors.

68. Which reactor uses heavy water as moderator?

Answer: CANDU.

69. Which reactor requires no moderator?

Answer: Breeder reactor.

70. Which reactor uses primary coolant as fluoride salts of lithium, beryllium, thorium and uranium?

Answer: Molten salt breeder reactor.

71. Why an increase in area is required to produce an increase of velocity in case of supersonic flow?

Answer: Increase in area for increase in velocity for supersonic flow is required because the density decreases faster than velocity increases at supersonic speeds and to maintain continuity of mass, area must increase.

72. Under what circumstances would there be an increase in pressure in a divergent nozzle?

Answer: For subsonic flow at inlet section of a diffuser a lower velocity and higher pressure will exist at the exit section. For supersonic isentropic flow at the inlet section a higher velocity and lower pressure will exist at the exit but if a shock wave occurs in the diffuser then a higher pressure will exist at the exit.

73. What are some conditions that may prevent a turbine from developing full power?
Answer:
1. The machine is overloaded.
2. The initial steam pressure and temperature are not up to design conditions.
3. The exhaust pressure is too high.
4. The governor is set too low.
5. The steam strainer is clogged.
6. Turbine nozzles are clogged with deposits.
7. Internal wear on nozzles and blades.

74. What is a stage in a steam turbine?
Answer: In an impulse turbine, the stage is a set of moving blades behind the nozzle. In a reaction turbine, each row of blades is called a "stage." A single Curtis stage may consist of two or more rows of moving blades.

75. What is a diaphragm?
Answer: Partitions between pressure stages in a turbine's casing are called diaphragms. They hold the vane-shaped nozzles and seals between the stages. Usually labyrinth-type seals are used. One-half of the diaphragms is fitted into the top of the casing, the other half into the bottom.

76. What are the two basic types of steam turbines?
Answer:
1. Impulse type.
2. Reaction type.

77. What are topping and superposed turbines?
Answer: Topping and superposed turbines arc high pressure, non-condensing units that can be added to an older, moderate-pressure plant. Topping turbines receive high-pressure steam from new high-pressure boilers. The exhaust steam of the new turbine has the same pressure as the old boilers and is used to supply the old turbines.

78. What is a combination thrust and radial bearing?
Answer: This unit has the ends of the Babbitt bearing extended radically over the end of the shell.
Collars on the rotor face these thrust pads, and the journal is supported in the bearing between the thrust collars.

79. What is important to remember about radial bearings?
Answer: A turbine rotor is supported by two radial bearings, one on each end of the steam cylinder. These bearings must be accurately aligned to maintain the close clearance between the shaft and the shaft seals, and between the rotor and the casing. If excessive bearing wear lowers the he rotor, great harm can be done to the turbine.

80. How many governors are needed for safe turbine operation? Why?

Answer: Two independent governors are needed for safe turbine operation. One is an over speed or emergency trip that shuts off the steam at 10 percent above running speed (maximum speed). The second, or main governor, usually controls speed at a constant rate; however, many applications have variable speed control.

81. How is a fly ball governor used with a hydraulic control?

Answer: As the turbine speeds up, the weights are moved outward by centrifugal force, causing linkage to open a pilot valve that admits and releases oil on either side of a piston or on one side of a spring-loaded piston. The movement of the piston controls the steam valves.

82. What is meant by critical speed?

Answer: It is the speed at which the machine vibrates most violently. It is due to many causes, such as imbalance or harmonic vibrations set up by the entire machine. To minimize damage, the turbine should be hurried through the known critical speed as rapidly as possible. (Caution, be sure the vibration is caused by critical speed and not by some other trouble).

83. How is oil pressure maintained when starting or stopping a medium-sized turbine?

Answer: An auxiliary pump is provided to maintain oil pressure. Some auxiliary pumps are turned by a hand crank; others are motor-driven. This pump is used when the integral pump is running too slowly to provide pressure, as when starting or securing a medium-sized turbine.

84. Besides lubrication, what are two functions of lubricating oil in some turbines?

Answer: In larger units, lube oil cools the bearings by carrying off heat to the oil coolers. Lube oil in some turbines also acts as a hydraulic fluid to operate the governor speed-control system.

85. In which part of the steam turbine does stress corrosion cracking (SCC) occur?

Answer: In the wet stages of the low-pressure turbine.

86. Do you stop cooling-water flow through a steam condenser as soon as the turbine is slopped?

Answer: You should keep the cooling water circulating for about 15 miles or more so that the condenser has a chance to cool down gradually and evenly. Be sure to have cooling water flowing through the condenser before starting up in order to prevent live steam from entering the condenser unless it is cooled. Overheating can cause severe leaks and other headaches.

87. How can the deposits be removed?
Answer:
1. Water soluble deposits may be washed off with condensate or wet steam.
2. Water insoluble deposits are removed mechanically after dismantling the turbine.
3. Experience shows that water soluble deposits are embedded in layers of water-insoluble deposits. And when the washing process is carried out, water soluble parts of the deposit dissolve away leaving a loose, friable skeleton of water insoluble deposits which then break loose and wash away.

88. How can the fatigue damage on high pressure blades be corrected?

Answer: Fatigue-damage on high-pressure blades arises due to vibration induced by partial-arc admission. This can be corrected by switching over to full arc admission technique.

89. How many governors are needed for safe turbine operation? Why?

Answer: Two independent governors are needed for safe turbine operation:
1. One is an over speed or emergency trip that shuts off the steam at 10 percent above running speed (maximum speed).
2. The second, or main governor, usually controls speed at a constant rate; however, many applications have variable speed control.

90. How will you detect that misalignment is the probable cause of excessive vibration?

Answer:
1. Coupling to the driven machine is to be disconnected.
2. The turbine is to be run alone.
3. If the turbine runs smoothly, misalignment, worn coupling or the driven equipment is the cause of the trouble.

91. In which part of the steam turbine does corrosion fatigue occur?

Answer: In the wet stages of the LP cylinder.

92. In which zone of steam turbines has temperature-creep rupture been observed?

Answer: Damage due to creep is encountered in high temperature (exceeding 455°C) zones. That is, it has been found to occur in the control stages of the high-pressure and intermediate-pressure turbines where steam temperature sometimes exceed 540°C. In the reheat stage, it has been observed that creep has caused complete lifting of the blade shroud bands.

93. What are the types of thrust bearings?

Answer:
1. Babbitt-faced collar bearings
2. Tilting pivotal pads
3. Tapered land bearings
4. Rolling-contact (roller or ball) bearings

94. What are the types of turbine seals?

Answer:
(A) Carbon rings fitted in segments around the shaft and held together by garter or retainer springs.
(B) Labyrinths mated with shaft serrations or shaft seal strips.
(C) Water seals where a shaft runner acts as a pump to create a ring of water around the shaft. Use only treated water to avoid shaft pitting.
(D) Stuffing box using woven or soft packing rings that are compressed with a gland to prevent leakage along the shaft.

95. What are the basic causes of the problem of rotor failure?

Answer:
- (A) Normal wear.
- (B) Fatigue failure due to high stress.
- (C) Design deficiency.
- (D) Aggressive operating environment

96. What are the differences between impulse and reaction turbines?

Answer:
- (A) The impulse turbine is characterized by the fact that it requires nozzles and that the pressures drop of steam takes place in the nozzles.
- (B) The reaction turbine, unlike the impulse turbines has no nozzles, as such. It consists of a row of blades mounted on a drum. The drum blades are separated by rows of fixed blades mounted in the turbine casing. These fixed blades serve as nozzles as well as the means of correcting the direction of steam onto the moving blades.
- (C) (C)In the case of reaction turbines, the pressure drop of steam takes place over the blades. This pressure drop produces a reaction and hence causes the motion of the rotor.

97. What is the function of EGR VALVE?

Answer: EGR means Exhaust Gas Recirculation. The main function of EGR is to control NOx emission from the engine. At elevated temperature (during long run & full condition)if due to some reasons engine gets heated up beyond certain level… it produces NOx emission. In order to avoid this pollution, we have to control the temperature rise inside the cylinder. This can be achieved by mixing some amount of exhaust gas with intake air. By doing this the temperature inside the combustion chamber is reduced. And NOx is controlled.

98. What is the difference between Critical speed & Whirling speed?

Answer: In Solid mechanics, in the field of rotor dynamics, the critical speed is the theoretical angular velocity which excites the natural frequency of a rotating object, such as a shaft, propeller or gear. As the speed of rotation approaches the objects natural frequency, the object begins to resonate which dramatically increases system vibration. The resulting resonance occurs regardless of orientation.
Whirling Speed is due to the unbalanced forces acting on a rotating shaft.

99. What is specific speed of turbine?

Answer: The specific speed is defined as the speed of the geometric turbine which can produce unit power at unit head

100. Material of Aircraft turbine blades

Answer: Ni, Cr, Al, Traces of C

Steam Turbine

01. What are the two basic types of steam turbines?
Answer:
1. Impulse type.
2. Reaction type.

02. How can steam turbines be classified?
Answer:
By the action of steam:
1. Impulse.
2. Reaction.
3. Impulse and reaction combined.
The number of step reductions involved:
4. Single stage.
5. Multi-stage.
6. Whether there is one or more revolving vanes separated by stationary reversing vanes.
The direction of steam flow:
7. Axial.
8. Radial.
9. Mixed.
10. Tangential.
11. Helical.
12. Re-entry.
The inlet steam pressure:
13. High pressure.
14. Medium pressure.
15. Low pressure.
The final pressure:
16. Condensing.
17. Non-condensing.
The source of steam:
18. Extraction.
19. Accumulator.

03. Do the radial axial-bore cracks occur in the LP rotor/shaft alone?
Answer: These are also known to occur in the HP as well as HP rotors.

04. Why should a steam or moisture separator be installed in the steam line next to a steam turbine?

Answer: All multistage turbines, low-pressure turbines, and turbines operating at high pressure with saturated steam should have a moisture separator in order to prevent rapid blade wear from water erosion.

05. Besides lubrication, which are two functions of lubricating oil in some turbines?

Answer: In large units, lube oil cools the bearings by carrying off heat to the oil coolers. Lube oil in some turbines also acts as a hydraulic fluid to operate the governor speed-control system.

06. Do you think that turbine blade failure is the only cause of unreliability of steam turbines? Does upgrading of turbine means replacement of blades and/or improvement of blade design?

Answer:

1. Like the blades, the steam-turbine rotors are highly stressed components. They are subject to cracking by a variety of failure mechanisms. Rotor failures do occur. And when they occur the result is catastrophic with the complete destruction of the unit and the total loss of generating capacity.
2. Therefore, special attention should be given to rotor upgrading and repairing techniques.

BLADE FAILURES

a. Unknown 26%
b. Stress-Corrosion Cracking 22%
c. High-Cycle Fatigue 20%
d. Corrosion-Fatigue Cracking 7%
e. Temperature Creep Rupture 6%
f. Low-Cycle Fatigue 5%
g. Corrosion 4%
h. Other causes 10%

Besides, many damage mechanisms operate in combination of, poor steam/water chemistry, certain blade design factors that vary from one turbine manufacture to other, system operating parameters,

07. How do the problems of vibration and fatigue arise with steam turbine blades?

Answer:

1. These arise due to flow irregularities introduced because of manufacturing defects, e.g. lack of control over tolerances.
2. System operating parameter, e.g. low flow may excite various modes of vibration in the blades.

08. How does the dirty safety trip valve trip the safety trip at normal speed? What is the remedy to it?

Answer: Dirt may find its way to the safety trip valve and get deposited around the spring end cap end. This will block the clearance between the safety trip valve and the spring end cap. As a result the steam pressure in the spring cap gets lowered allowing the valve to close.

The remedy is:

The spring end cap as well as safety trip valve should be cleaned.

09. What maybe the possible causes for the safety trip tripping during load variation?

Answer:

1. Light load and high inlet steam pressure.
2. Safety trip set very close to the operating speed of turbine.

10. What is the safe maximum tripping speed of a turbine operating at 2500 rpm?

Answer: The rule is to trip at 10 percent over speed. Therefore, 2500 x 1.10 = 2750 rpm

11. What is the purpose of a turning gear?

Answer: Heat must be prevented from warping the rotors of large turbines or high-temperature turbines of 400°C or more. When the turbine is being shut down, a motor-driven turning gear is engaged to the turbine to rotate the spindle and allow uniform cooling.

12. What is the potential problem of shrunk-on-disc type rotor?

Answer:

1. It is the failure due to circumferential cracks, which are not limited to old rotors of early models (1960), but they also take place on present-day rotors.
2. As a result corrodents' impurities like chlorides concentrate at key ways. This factor coupled with high stress concentration lead to SCC attack on keyway areas.

13. What is the principle of a steam turbine?

Answer:

1. If high-velocity steam is allowed to blow on to a curved blade, the steam will suffer a change in direction as it passes across the blade.
2. As a result of its change in direction across the blade, the steam will impart a force to the blade.
3. Now if the blade were free in the direction of force as depicted. If, therefore, a number of blades were fixed on the circumference of a disc which is free to rotate on a shaft, then steam blown across the blades in the way described, would cause the disc to rotate. This is the working principle of a steam turbine.

14. What is the function of a gland drain?

Answer: The function of a gland drain is to draw off water from sealing-gland cavities created by the condensation of the sealing steam. Drains are led to either the condenser air-ejector tube nest or the feed water heaters. Often, gland drains are led to a low-pressure stage of the turbine to extract more work from the gland-sealing steam.

15. What is the difference between partial and full arc admission?
Answer: In multi-valve turbine inlets, partial arc admission allows the steam to enter per valve opening in a sequential manner, so as load is increased, more valves open to admit steam. This can cause uneven heating on the high-pressure annulus as the valves are individually opened with load increase. In full-arc admission, all regulating valves open but only at a percentage of their full opening. With load increase, they all open more fully. This provides more uniform heating around the high-pressure part of the turbine. Most modern controls start with full-arc and switch to partial arc to reduce throttling losses through the valves.

16. What is the cause of circumferential cracking?
Answer: High cycle fatigue with or without corrosion.

17. What is meant by the water rate of a turbine?
Answer: It is the amount of water (steam) used by the turbine in pounds per horsepower per hour or kilowatts per hour.

18. What is gland-sealing steam?
Answer: Low-pressure steam is led to a sealing gland. The steam seals the gland, which may be a carbon ring or of the labyrinth type, against air at the vacuum end of the shaft.

19. What is meant by critical speed?
Answer: It is the speed at, which the machine vibrates most violently. It is due to many causes, such as imbalance or harmonic vibrations set up by the entire machine. To minimize damage, the turbine should be hurried through the known critical speed as rapidly as possible. Be sure the vibration is caused by critical speed and not by some other trouble.

20. How does the damage to turbine-blades tell upon the efficiency of the unit?
Answer: The damage to blade profiles changes the geometry of steam flow path and thereby reducing the efficiency of the unit.

21. What is the possible cause of slow start up of a steam turbine?
Answer: This may be due to high starting torque required by the driven equipment.

22. What is the remedy for rotor-surface cracking?
Answer: Current rotor/shaft should be machined off (skin-peeling).

23. What is to be done in case of cracks originating at the lacing-wire holes?
Answer: These are to be weld-repaired. However the following factors must be considered:
   a.   The length of the crack that appears on the pressure and/or suction face.

b. Whether the cracks propagate towards inlet end, discharge end and or both.

24. What is the essential distinguishing feature between a steam turbine and reciprocating steam engine?
Answer:
1. In a steam turbine, the heat energy of steam is converted into kinetic energy by allowing it to expand through a series of nozzles and this kinetic energy of steam is then imparted to the turbine blades mounted on a shaft free to rotate to drive this prime mover.
2. In a reciprocating steam engine, the pressure energy of steam is directly utilized to overcome the external resistance. Here, the utilization of the KE of input steam is negligibly small.

25. How can the problem of excessive speed variation due to throttle assembly friction be overcome?
Answer: The throttle should be dismantled. Moving parts should be checked for free and smooth movement. Using very fine-grained emery paper, the throttle valve seats and valve steam should be polished.

26. How can damaged tenons be repaired?
Answer: By adopting modern welding techniques, tenons can be rebuilt This in some cases results in extended blade life.

27. How can the fatigue damage on high-pressure blades be corrected?
Answer: Fatigue-damage on high-pressure blades arises due to vibration induced by partial-arc admission. This can be corrected by switching over to full arc admission technique.

28. In which turbines, is this pressure-velocity compounding principle employed?
Answer: In the Curtis turbine.

29. How is oil pressure maintained when starting or stopping a medium-sized turbine?
Answer: An auxiliary pump is provided to maintain oil pressure. Some auxiliary pumps are turned by a hand crank; others are motor-driven. This pump is used when the integral pump is running too slowly to provide pressure, as when starting or securing a medium-sized turbine.

30. In which cases does moisture-impingement and washing erosion occur?
Answer:
1. These are encountered in the wet sections of the steam turbine.
2. For nuclear power plants, these wet sections can involve parts of high pressure cylinder.

31. In which section of the steam-turbine rotors is the problem of rotor failure mostly prevalent?
Answer: Rotor failures occur mostly on the large low-pressure rotors.

32. How does pressure monitoring ensure detection of turbine deposits?
Answer:
1. Pressure of steam expanding in the turbine is measured at characteristic points, i.e., at the wheel chamber, points of pass-out, inlet/outlet of HP, IP and LP stages of the turbine.
2. The turbine manufacturer provides the pressure characteristics in the form of graphs.
3. At 1st commissioning, the user supplements these theoretical curves with those derived from actual measurements. These are actual pressure characteristics for a clean turbine. Now these pressure characteristics are compared with those obtained during operation in the later period.
4. Under identical conditions, an increase in pressure shows the formation of deposits.
5. For a steam throughput in the range 70-100%, an increase in wheel chamber pressure of more than 10% indicates severe blade depositions.

33. What are the disadvantages of velocity compounding?
Answer:
1. Steam velocity is too high and that is responsible for appreciable friction losses.
2. Blade efficiency decreases with the increase of the number of stages.
3. With the increase of the number of rows, the power developed in successive rows of blade decreases. For as much as, the same space and material are required for each stage, it means, therefore, that all stages are not economically efficient.

34. What are the main causes of turbine vibration?
Answer:
1. Unbalanced parts
2. Poor alignment of parts
3. Loose parts
4. Rubbing parts
5. Lubrication troubles
6. Steam troubles
7. Foundation troubles
8. Cracked or excessively worn parts

35. What is an air ejector?
Answer: An air ejector is a steam siphon that removes non-condensable gases from the condenser.

36. What are the possible causes of a governor not operating?
Answer:
1. Restriction of throttle valve reflex.
2. Failure of governor control on start-up.

If it is found that after start-up, the speed increases continuously and the governor is not closing the throttle valve, it may be that the governor pump has been installed in the wrong direction.

37. What are two types of clearance in a turbine?
Answer:
1. Radial - the clearance at the tips of the rotor and casing
2. Axial - the fore-and-aft clearance, at the sides of the rotor and the casing

38. What is a combination thrust and radial bearing?
Answer:
This unit has the ends of the Babbitt bearing extended radially over the end of the shell. Collars on the rotor face these thrust pads, and the journal is supported in the bearing between the thrust collars.

39. What is the nature of circumferential cracking in shrunk-on-disc rotors in steam turbines?
Answer: Regions of high stress concentration give birth to this type of cracking. It begins in corrosion pits and propagates towards the bore by high-cycle fatigue. It may culminate in a catastrophe, if it penetrates the bore (happily this usually does not occur).

40. What is the function of a thrust bearing?
Answer: Thrust bearings keep the rotor in its correct axial position.

41. What is to be done for erosion-induced damage on high-and low-pressure stage blading?
Answer:
1. In such cases welding repair can be a good solution and this can be carried out during a normal maintenance outage without removing the blade. Using oxyacetylene torch, Satellites is generally deposited onto the damaged site. Following this, the weld is subjected to stress-relieving and re-profiling.
2. In case of erosion penetrating the erosion shield and extending to the base material, a filler material of consistent or identical composition of blade material is used.
3. In some cases use is made of Inconel alloy to build up the metal base. Therefore, using welding or brazing technique, a new shield can be attached to the blade. If brazing technique is followed, the rebuilt section is stress relieved prior to the attachment of shield to it. If, on the other hand, the shield is attached by welding, then they are stress-relieved together.

42. What is done when cracks due to SCC or corrosion-fatigue are found?
Answer: The damaged blade is usually replaced, as repairing is difficult.

43. What is a stage in a steam turbine?

Answer: In an impulse turbine, the stage is a set of moving blades behind the nozzle. In a reaction turbine, each row of blades is called a "stage." A single Curtis stage may consist of two or more rows of moving blades.

44. How does deposit formation on turbine blades affect turbine efficiency?

Answer: About 500 g of deposits distributed more or less evenly all over the blading section can bring down turbine efficiency by 1%.

45. How does the internal efficiency monitoring lead to the detection of turbine deposits?

Answer:
1. Process heat drop.
2. Adiabatic heat drop.
3. The process heat drop and adiabatic heat drop are obtained from a Mollier- Chart for the corresponding values of steam parameters - pressure and temperature - at initial and final conditions.

46. What are the possible causes of the speed of the turbine rotor increasing excessively as the load is decreased?

Answer:
1. Throttle valve not closing fully.
2. Wearing of throttle valve seats.

47. What are the points of SCC attack? What are these points in particular?

Answer: SCC attack predominates where corrodents deposit and build up i.e. in those blading areas where flowing steam cannot provide a washing effect.

These points are:
a. Tie wires.
b. Tie wire holes.
c. Brazing.
d. Blade covers.
e. Tenons holes.

48. What are the causes of radial axial-bore cracks on HP/IP rotors/shafts?

Answer:
1. The predominant cause is creep, which may act with or without low cycle fatigue.
2. Also the cracks result due to poor creep ductility due to faulty heat treatment process.

49. What is the operating principle of an impulse turbine?

Answer: The basic idea of an impulse turbine is that a jet of steam from a fixed nozzle pushes against the rotor blades and impels them forward. The velocity of the steam is about twice as fast as the velocity of the blades. Only turbines utilizing fixed nozzles are classified as impulse turbines.

50. Despite preventive measures, damage due to moisture impingement has been found, in certain cases, in the shield and beyond. Why?
Answer:
1. Shields are designed and fabricated on the basis of predicted range of steam/water quantities impacting the blades at specific angles.
2. Now if the operating conditions deviate significantly from design parameters then the erosion damage will occur. And in some cases it may go beyond nominal erosion wear and warrant repair.
3. Also the corrosion of casing can occur due to blockage/clogging of water drains or extraction thereby forcing the water back into the casing. If this condensate water is carried over to steam path and impacts the blade, thermal-fatigue failure can occur within a short period.

51.In which turbine tip leakage is a problem?
Answer: Tip leakage is a problem in reaction turbines. Here, each vane forms a nozzle; steam must flow through the moving nozzle to the fixed nozzle. Steam escaping across the tips of the blades represents a loss of work. Therefore, tip seals are used to prevent this.

52. What are four types of thrust hearings?
Answer:
1. Babbitt-faced collar bearings.
2. Tilting pivotal pads.
3. Tapered land bearings.
4. Rolling-contact (roller or ball) bearings.

53. Why is it necessary to open casing drains and drains on the steam line going to the turbine when a turbine is to be started?
Answer: To avoid slugging nozzles and blades inside the turbine with condensate on start-up; this can break these components from impact. The blades were designed to handle steam, not water.

54. What are some conditions that may prevent a turbine from developing full power?
Answer:
1. The machine is overloaded.
2. The initial steam pressure and temperature are not up to design conditions.
3. The exhaust pressure is too high.
4. The governor is set too low.
5. The steam strainer is clogged.
6. Turbine nozzles are clogged with deposits.

7. Internal wear on nozzles and blades.

55. What steam rate is as applied to turbo-generators?
Answer: The steam rate is the pounds of steam that must be supplied per kilowatt-hour of generator output at the steam turbine inlet.

56. What is the operating principle of a reaction turbine?
Answer: A reaction turbine utilizes a jet of steam that flows from a nozzle on the rotor. Actually, the steam is directed into the moving blades by fixed blades designed to expand the steam. The result is a small increase in velocity over that of the moving blades. These blades form a wall of moving nozzles that further expand the steam. The steam flow is partially reversed by the moving blades, producing a reaction on the blades. Since the pressure drop is small across each row of nozzles (blades), the speed is comparatively low. Therefore, more rows of moving blades are needed than in an impulse turbine.

57. What is a multi-port governor valve? Why is it used?
Answer: In large turbines, a valve controls steam flow to groups of nozzles. The number of open valves controls the number of nozzles in use according to the load. A bar-lift or cam arrangement operated by the governor opens and closes these valves in sequence. Such a device is a multi-port valve. Using nozzles at full steam pressure is more efficient than throttling the steam.

58. Besides lubrication, what are two functions of lubricating oil in some turbines?
Answer: In larger units, lube oil cools the bearings by carrying off heat to the oil coolers. Lube oil in some turbines also acts as a hydraulic fluid to operate the governor speed-control system.

59. By monitoring the exhaust steam temperature, how can the blade deposition be predicted?
Answer:
1. Immediately after the 1st commissioning, the different values of exhaust temperature for different steam flow rates are precisely determined and plotted against steam flow. This will produce the first actual graph. This is for a clean turbine.
2. Similar graphs are to be drawn at later periods for comparing with the initial graph.
3. A rise in exhaust steam temperature under the same conditions refers to deposit formation.
4. An increase of exhaust steam temperature by more than 10% in the range of 70 to 100% steam flow indicates inadmissible blade depositions. Shutdown is to be taken and blades are to be washed off deposits.

60. Do you stop cooling-water flow through a steam condenser as soon as the turbine is slopped?

Answer: You should keep the cooling water circulating for about 15 mills or more so that the condenser has a chance to cool down gradually and evenly. Be sure to have cooling water flowing through the condenser before starting up in order to prevent live steam from entering the condenser unless it is cooled. Overheating can cause severe leaks and other headaches.

61. How can problems of "excessive vibration or noise" due to piping strain be avoided on steam turbines?

Answer:
1. The inlet as well as exhaust steam lines should be firmly supported to avoid strains from being imposed on the turbine.
2. Adequate allowance should be made for expansion of steam pipes due to heat.

62. How can the detection of deposits in a turbine be made during operation?

Answer:
1. Pressure monitoring.
2. Internal efficiency monitoring.
3. Monitoring exhaust steam temperature.
4. Monitoring specific steam consumption.

63. How can the disadvantages of the impulse turbine be overcome?

Answer:
1. By Velocity compounding
2. By Pressure compounding
3. By Pressure-Velocity compounding.

64. How can the misalignment be rectified?

Answer: The bolts holding the flanges together are to be tightened. The coupling is to be checked for square between the bore and the face. At the same time axial clearance is to be checked. Using gauge block and feeler gauges, the gap between coupling faces 1800 apart is to be measured. After rotating the coupling-half 1800, the gap at the same points is to be measured. After this, the other coupling is to be rotated 1800 and the gap at the same points is to be re-measured. These measures should come within a few thousands of an inch. Dividing the coupling faces into four intervals, the distance between the coupling faces at this intervals is to be measured with the aid of a gauge block and feeler gauges. These gaps measurements should come within 0.005 inch for proper angular shaft alignment. After proper alignment at room temperature, the two halves of the coupling are to be connected.

65. How can the speed variation be reduced by making a governor droop adjustment?

Answer: If the internal droop setting is increased, the speed variation will reduce.

66. How does improper governor lubrication affect? What is the remedy to it?

Answer: In the event of low governor oil level or if the oil is dirty or foamy, it will cause improper governor lubrication.

The remedy is:

1. The dirty or foamy lube oil should be drained off; governor should be flushed and refilled with a fresh charge of proper oil.
2. In the event of low level, the level should be built up by make- up lube oil.

67. How does solid-particle erosion occur?

Answer: Solid-particle erosion, i.e. SPE occurs in the high-pressure blades. And it takes place when hard particles of iron exfoliated by steam from superheater tubes, reheater tubes, steam headers and steam leads strike on the surface of turbine blades.

68. How does the foreign-particle damage of turbine blades arise?

Answer: It occurs due to impact on blades by foreign particles (debris) left in the system following outages and become steam-borne later.

69. How does this modification reduce the vibration fatigue damage?

Answer:

1. Joining the blade segments together at the shroud band increases the length of the arc-to a maximum of 360° that alters the natural frequency of the blade grouping from the operating vibration mode.
2. This design has gained considerable success in commercial service.

70. How is a flyball governor used with a hydraulic control?

Answer: As the turbine speeds up, the weights are moved outward by centrifugal force, causing linkage to open a pilot valve that admits and releases oil on either side of a piston or on one side of a spring-loaded piston. The movement of the piston controls the steam valves.

71. How is pressure compounding accomplished?

Answer:

1. This is accomplished by an arrangement with alternate rows of nozzles and moving blades.
2. Steam enters the 1st row of nozzles where it suffers a partial drop of pressure and in lieu of that its velocity gets increased. The high velocity steam passes on to the 1st row of moving blades where its velocity is reduced.
3. The steam then passes into the 2nd row of nozzles where its pressure is again partially reduced and velocity is again increased. This high velocity steam passes from the nozzles to the 2nd row of blades where its velocity is again reduced.
4. Thus pressure drop takes place in successive stages. Since a partial pressure drop takes place in each stage, the steam velocities will not be so high with the effect that the turbine will run slower.

72. How is pressure-velocity compounding accomplished?
Answer:
1. It is a combination of pressure compounding and velocity compounding.
2. Steam is expanded partially in a row of nozzles whereupon its velocity gets increased. This high velocity steam then enters a few rows of velocity compounding whereupon its velocity gets successively reduced.
3. The velocity of the steam is again increased in the subsequent row of nozzles and then again it is allowed to pass onto another set of velocity compounding that brings about a stage-wise reduction of velocity of the steam.
4. This system is continued.

73. How is the washing of turbine blades carried out with the condensate?
Answer:
1. The washing is carried out with the condensate at 100°C.
2. The turbine is cooled or heated up to 100°C and filled with the condensate via a turbine drain.
3. The rotor is turned or barred by hand and the condensate is drained after 2 to 4 hours.
4. It is then again filled with the condensate at 100°C (but up to the rotor centre level), the rotor is rotated and the condensate is drained after sometime. This process is repeated several times.

74. How is turbine blade washing with wet steam carried out?
Answer:
1. Wet steam produced usually by injecting cold condensate into the superheated steam, is introduced to the turbine which is kept on running at about 20% of nominal speed.
2. For backpressure turbine the exhaust steam is let out into the open air through a gate valve. For a condensing turbine, the vacuum pump is kept out of service while cooling water is running, with the effect that the entering cooling steam is condensed. The condensate is drained off.
3. The washing steam condition is gradually adjusted to a final wetness of 0.9 to 0.95.

Note, it is important
4. Not to change washing steam temperature by 10°C/min,
5. To keep all turbine cylinder drains open.

75. How is velocity compounding accomplished?
Answer:
1. This is accomplished by an arrangement with alternate rows of fixed blades and moving blades. They mounted on the casing while the moving blades are keyed in series on a common shaft. The function of the fixed blades is to correct the direction of entry of steam to the next row of moving blades.
2. The high velocity steam leaving the nozzles passes on to the 1st row of moving blades where it suffers a partial velocity drop.

3. Its direction is then corrected by the next row of fixed blades and then it enters the 2nd row of moving blades. Here the steam velocity is again partially reduced. Since only part of the velocity of the steam is used up in each row of the moving blades, a slower turbine results. This is how velocity compounding works.

76. How many governors are needed for safe turbine operation? Why?

Answer: Two independent governors are needed for safe turbine operation:
1. One is an over speed or emergency trip that shuts off the steam at 10 percent above running speed (maximum speed).
2. The second, or main governor, usually controls speed at a constant rate; however, many applications have variable speed control.

77. How many types of particle-impact damage occur in turbine blades?

Answer:
1. Erosion/corrosion.
2. Foreign-particle impacts.
3. Solid-particle erosion.
4. Water damage.

78. How to prevent turbine deposition?

Answer: By upgrading the quality of steam and by ensuring proper quality of the following.
1. Boiler feed water quality.
2. Steam boiler model.
3. Boiler design.
4. Boiler operation.

79. How will you detect that misalignment is the probable cause of excessive vibration?

Answer:
1. Coupling to the driven machine is to be disconnected.
2. The turbine is to be run alone.
3. If the turbine runs smoothly, misalignment, worn coupling or the driven equipment is the cause of the trouble.

80. How would you slop a leaky tube in a condenser that was contaminating the feed water?

Answer: To stop a leaky tube from contaminating the feed water, shut down, remove the water-box covers, and fill the steam space with water. By observing the tube ends you can find the leaky tube. An alternate method is to put a few pounds of air pressure in the steam space, flood the water boxes to the top inspection plate, and observe any air bubbles. Once you have found the leaky tube, drive a tapered bronze plug (coated with white lead) into each end of the tube to cut it out of service. This allows you to use the condenser since the tubes need not be renewed until about 10 percent of the tubes are plugged.

81. In which part of the steam turbine does stress corrosion cracking (SCC) occur?
Answer: In the wet stages of the low-pressure turbine.

82. In how many patterns are tie wires used?
Answer:
1. In one design, tie wire is passed through the blade vane.
2. In another design, an integral stub is joined by welding/brazing.

83. In steam turbines, is there any alternative to the shrunk-on-disc design?
Answer: Two designs are available at present:
1. Welded rotor in which each individual disc is welded, instead of shrunk, onto the main shaft.
2. Monobloc rotor in which the entire shaft and blade assembly is manufactured from a single forging.

84. In which case does upgrading imply life extension of steam turbines?
Answer: For a capital-short electric utility plant, upgrading comes to mean extending the life of that plant scheduled for retirement.

85. In which cases does erosion corrosion damage appear?
Answer: It is commonly encountered in nuclear steam turbines and old fossil-fuel-fired units that employ lower steam temperatures and pressures.

86. In which cases does upgrading mean up-rating the turbine capacity?
Answer: For an electric utility system facing uncertain load growth, upgrading is chiefly up rating.
It is an inexpensive way to add capacity in small increments.

87. In which part of the steam turbine does corrosion fatigue occur?
Answer: In the wet stages of the LP cylinder.

88. How would you stop air from leaking into a condenser?
Answer: First, find the leak by passing a flame over the suspected part while the condenser is under vacuum. Leaks in the flange joints or porous castings can be stopped with asphalt paint or shellac. Tallow or heavy grease will stop leaks around the valve stems. Small leaks around the porous castings, flange nuts, or valve stems can always be found by the flame test. So, you might have to put the condenser under a few pounds of air pressure and apply soapsuds to the suspected trouble parts.

89. What are the basic causes of the problems are?
Answer:
1. Normal wear.

2. Fatigue failure due to high stress.
3. Design deficiency.
4. Aggressive operating environment

90. In which turbine is this pressure compounding used?
Answer: In the Rateau turbine.

91. In which turbine is velocity compounding utilized?
Answer: In the Curtis turbine.

92. In which zone of steam turbines has temperature-creep rupture been observed?
Answer: Damage due to creep is encountered in high temperature (exceeding 455°C) zones. That is, it has been found to occur in the control stages of the high-pressure and intermediate-pressure turbines where steam temperature sometimes exceed 540°C. In the reheat stage, it has been observed that creep has caused complete lifting of the blade shroud bands.

93. Is there any adverse effect off full-arc admission operation?
Answer: At low loads, this results in a heat-rate penalty, due to throttling over the admission valves.

94. Is there any other type of racking occurring in HP/IP rotors and causing rotor failures?
Answer:
1. Blade-groove-wall cracking.
2. Rotor-surface cracking.

95. Of all the factors that contribute to the unreliability of steam turbines, which one is the most prominent?
Answer: It is the problem of turbine blade failures that chiefly contribute to the unreliability of steam turbines.

96. Rim cracking continues to be a problem of shrunk-on-disc type rotors in utility steam turbines. Where does it occur?
Answer: Cracking has been located at the outer corners of tile grooves where the blade root attaches to the rotor.

97. So can you recommend this technique as a permanent measure?
Answer: No, this can be recommended in extreme cases or at best temporarily.

98. What should be the more sound approach in the case of steam turbine blade failure?
Answer: The more reasonable and better approach is to replace the damaged blades with new ones that are stiffened by:

a. Secreting the interface surface of individual blades so they interlock, or
b. Welding the blades together.
c. In some cases, a single monolithic block is machined out to manufacture the blades in a group.
d. In some other cases, blades themselves are directly welded into the rotor.

99. Steam blowing from a turbine gland is wasteful. Why else should it be avoided?
Answer: It should be avoided because the steam usually blows into the bearing, destroying the lube oil in the main bearing. Steam blowing from a turbine gland also creates condensate, causing undue moisture in plant equipment.

100. What are the consequences of turbine depositions?
Answer: The consequences of turbine depositions have three effects.
1. Economic Effect:
a. Reduction in turbine output
b. Decrease in efficiency requiring higher steam consumption.

2. Effect of Overloading and Decreasing Reliability in Operation:
a. Pressure characteristic in the turbine gets disturbed with the effect that thrust and overloading of thrust bearing increase.
b. Blades are subjected to higher bending stresses.
c. Natural vibrations of the blading are affected.
d. Vibration due to uneven deposition on turbine blading.
e. Valve jamming due to deposits on valve stems.

3. Corrosion Effect:
a. Fatigue corrosion
b. Pitting corrosion.
c. Stress corrosion.

101. Usually it has been found that SCC attack takes place particularly at key-ways of shrunk-on-disc rotors of low-pressure turbines. Why are keyways prone to SCC attack?
Answer:
1. Keyways shrunk-fit each disc onto tile rotor shaft. They improve the rigidity of the connection between the disc and the central shaft However, key ways are subjected to abnormal centrifugal forces due to high over speed, that reduce the amount of shrink. Tangential stresses tend to gravitate at the keyway connection and steam tends to condense.
2. It is a one-piece-construction, and thus has inherent rigidity.
3. Advanced steel making techniques enable building of mono block rotors almost free from non-metallic inclusions and gas bubbles. Even large mono block rotors of clean steel are being manufactured today.

4. It exhibits lower inherent stresses.
5. The chance of disc loosening during operation is eliminated.
6. Highly stressed keyway is eliminated.

102. What are four types of thrust bearings?
Answer:
1. Babbitt-faced collar bearings
2. Tilting pivotal pads
3. Tapered land bearings
4. Rolling-contact (roller or ball) bearings

103. What are four types of turbine seals?
Answer:
1. Carbon rings fitted in segments around the shaft and held together by garter or retainer springs.
2. Labyrinths mated with shaft serrations or shaft seal strips.
3. Water seals where a shaft runner acts as a pump to create a ring of water around the shaft. Use only treated water to avoid shaft pitting.
4. Stuffing box using woven or soft packing rings that are compressed with a gland to prevent leakage along the shaft.

104. What are some common troubles in surface-condenser operation?
Answer:
The greatest headache to the operator is loss of vacuum caused by air leaking into the surface condenser through the joints or packing glands. Another trouble spot is cooling water leaking into the steam space through the ends of the tubes or through tiny holes in the tubes. The tubes may also become plugged with mud, shells, debris, slime, or algae, thus cutting down on the cooling-water supply or the tubes may get coated with lube oil from the reciprocating machinery. Corrosion and dezincification of the tube metal are common surface condenser troubles. Corrosion may be uniform, or it may occur in small holes or pits. Dezincification changes the nature of the metal and causes it to become brittle and weak.

105. What are the advantages of steam turbines over reciprocating steam engines?
Answer:
1. Steam turbine has higher thermal efficiency than reciprocating steam engines.
2. The brake horsepower of steam turbines can range from a few HP to several hundred thousand HP in single units. Hence they are quite suitable for large thermal power stations.
3. Unlike reciprocating engines, the turbines do not need any flywheel, as the power delivered by the turbine is uniform.
4. Steam turbines are perfectly balanced and hence present minimum vibrational problem.

5. High rpm 18000 - 24000 can be developed in steam turbines but such a high speed generation is not possible in the case of reciprocating steam engines.
6. Some amount of input energy of steam is lost as the reciprocating motion of the piston is converted to circular motion.
7. Unlike reciprocating steam engines, no internal lubrication is required for steam turbines due to the absence of rubbing parts.
8. Steam turbines, if well designed and properly maintained, are more reliable and durable prime movers than steam engines.

106. What are the advantages of velocity compounding?
Answer:
1. The velocity compounding system is easy to operate and operation is more reliable.
2. Only two or three stages are required. Therefore, first cost is less.
3. Since the total pressure drop takes place only in nozzles and not in the blades, the turbine casing need not be heavily built. Hence, the economy in material and money.
4. Less floor space is required.

107. What are the advantages of welded rotors?
Answer:
1. Welded rotor is a composed body built up by welding the individual segments. So the limitations on forgings capacity do not apply.
2. Welding discs together results in a lower stress level. Therefore, more ductile materials can be chosen to resist SCC attack.
3. There are no keyways. So regions of high stress concentrations are eliminated.

108. What are the basic causes of the problem of rotor failure?
Answer:
1. Normal wear.
2. Fatigue failure due to high stress.
3. Design deficiency.
4. Aggressive operating environment

109. What are the differences between impulse and reaction turbines?
Answer:
1. The impulse turbine is characterized by the fact that it requires nozzles and that the pressure drops of steam takes place in the nozzles.
2. The reaction turbine, unlike the impulse turbines has no nozzles, as such. It consists of a row of blades mounted on a drum. The drum blades are separated by rows of fixed blades mounted in the turbine casing. These fixed blades serve as nozzles as well as the means of correcting the direction of steam onto the moving blades.
3. In the case of reaction turbines, the pressure drop of steam takes place over the blades. This pressure drop produces a reaction and hence causes the motion of the rotor.

110. What are the factors that contribute to bearing failure in a steam turbine?

Answer:

1. Improper lubrication. Only the recommended lubricant should be used.
2. Inadequate water-cooling.
    (a) The jacket temperature should be maintained in the range of 37-60°C
    (b) The flow of cooling water should be adjusted accordingly.
3. Misalignment: It is desirable that ball bearings should fit on the turbine shaft with a light press fit. If the fitting is too tight, it will cause cramping. On the other hand, if the fitting is too loose it will cause the inner race to turn on the shaft. Both conditions are undesirable. They result in wear, excessive vibration and overheating. And bearing failure becomes the ultimate result.
4. Bearing fit.
5. Excessive thrust.
6. Unbalance.
7. Rusting of bearing.

111. What are the losses in steam turbines?

Answer:

1. Residual Velocity Loss - This is equal to the absolute velocity of the steam at the blade exit.
2. Loss due to Friction - Friction loss occurs in the nozzles, turbine blades and between the steam and rotating discs. This loss is about 10%.
3. Leakage Loss.
4. Loss due to Mechanical Friction - Accounts for the loss due to friction between the shaft and bearing.
5. Radiation Loss - Though this loss is negligible, as turbine casings are insulated, it occurs due to heat leakage from turbine to ambient air which is at a much lower temperature than the turbine.
6. Loss due to Moisture - In the lower stages of the turbine, the steam may become wet as the velocity of water particles is lower than that of steam. So a part of the kinetic energy of steam is lost to drag the water particles along with it.

112. At what points does corrosion fatigue does show up?

Answer: It attacks trailing edges, near the base of the foil and also the blade-root serration's.

113. What are the possible causes for the turbine not running at rated speed?

Answer: The possible causes are:

1. Too many hand valves closed,
2. Oil relay governor set too low,
3. Inlet steam pressure too low or exhaust pressure too high,
4. Load higher than turbine rating,
5. Throttle valve not opening fully,

6. Safety trip valve not opening properly,
7. Nozzles plugged,
8. Steam strainer choked.

114. What are the possible causes of excessive vibration or noise in a steam turbine?
Answer:
1. Misalignment.
2. Worn bearings.
3. Worn coupling to driven machine.
4. Unbalanced coupling to driven machine.
5. Unbalanced wheel.
6. Piping strain.
7. Bent shaft.

115. What are the stresses to which a steam turbine rotor is subjected during its service life?
Answer:
1. **Mechanical stress** - The factors that contribute to mechanical stress in the shaft are the centrifugal forces and torque's generated due to revolving motion of the shaft as well as bending arising during steady-state operation.
2. **Thermal stress** - Transient operating phases i.e. start-up and shutdown the genesis of thermal stress induced to the turbine shaft.
3. **Electrically induced stress** - They originate due to short circuits and faulty synchronization.

116. What are three types of condensers?
Answer:
1. Surface (shell-and-tube)
2. Jet
3. Barometric.

117. What are topping and superposed turbines?
Answer:
Topping and superposed turbines are high-pressure, non-condensing units that can be added to an older, moderate-pressure plant. Topping turbines receive high-pressure steam from new high-pressure boilers. The exhaust steam of the new turbine is at the same pressure as the old boilers and is used to supply the old turbines.

118. What design modification is adopted to reduce susceptibility of last low pressure stages to fatigue failure?
Answer: One modification is to join the blade segments together at the shroud band.

119. What does "upgrading" generally means in the context of steam turbines?

Answer: Upgrading is a most widely used tern. It encompasses a variety of meanings verses life extension, modernization and up-rating of steam turbines.

120. What does the term "ramp rat" mean?

Answer: Ramp rate is used in bringing a turbine up to operating temperature and is the degrees Fahrenheit rise per hour that metal surfaces are exposed to when bringing a machine to rated conditions. Manufactures specify ramp rates for their machines in order to avoid thermal stresses. Thermocouples are used in measuring metal temperatures.

121. What factors are responsible for turbine-blade failures?

Answer:

In the high pressure cylinder, the turbine blades are mostly affected by:

1. Solid-particle erosion (SPE)
2. High cycle fatigue

Whereas, in the last few stages of the low-pressure cylinder, the blade damage is mainly afflicted by:

1. Erosion
2. Corrosion
3. Stress/fatigue damage mechanism
4. According to EPRI (Electric Power Research Institute, USA) data stress corrosion cracking and fatigue are the chief exponents for turbine-blade failures in utility industries.

122. What factors cause excessive steam leakage under carbon rings?

Answer:

1. Dirt under rings: steam borne scale or dirt foul up the rings if steam is leaking under the carbon rings.
2. Shaft scored.
3. Worn or broken carbon rings. These should be replaced with a new set of carbon rings. The complete ring is to be replaced.

123. What factors contribute to excessive speed variation of the turbine?

Answer:

1. Improper governor droop adjustment.
2. Improper governor lubrication.
3. Throttle assembly friction.
4. Friction in stuffing box.
5. High inlet steam pressure and light load.
6. Rapidly varying load.

124. What is a balance piston?

Answer: Reaction turbines have axial thrust because pressure on the entering side is greater than pressure on the leaving side of each stage. To counteract this force, steam is admitted to a dummy (balance) piston chamber at the low-pressure end of the rotor. Some designers also use a balance piston on impulse turbines that have a high thrust. Instead of pistons, seal strips are also used to duplicate a piston's counter force.

125. What is a diaphragm (turbine)?

Answer: Partitions between pressure stages in a turbine's casing are called diaphragms. They hold the vane-shaped nozzles and seals between the stages. Usually labyrinth-type seals are used.

One-half of the diaphragms are fitted into the top of the casing, the other half into the bottom.

126. What is a multiport governor valve? Why is it used?

Answer: In large turbines, a valve controls steam flow to groups of nozzles. The number of open valves controls the number of nozzles in use according to the load. A bar-lift or cam arrangement operated by the governor, opens and closes the valves in sequence. Such a device is a multiport valve. Using nozzles at full steam pressure is more efficient than throttling the steam.

127. What is a radial-flow turbine?

Answer: In a radial-flow turbine, steam flows outward from the shaft to the casing. The unit is usually a reaction unit, having both fixed and moving blades. They are used for special jobs and are more common to European manufacturers.

128. What is a shrunk-on-disc rotor?

Answer: These are built by heat expanding the discs, so that upon cooling they shrink on the main rotor forging.

129. What is a tapered-land thrust bearing?

Answer: The Babbitt face of a tapered-land thrust bearing has a series of fixed pads divided by radial slots. The leading edge of each sector is tapered, allowing an oil wedge to build up and carry the thrust between the collar and pad.

130. What is an extraction turbine?

Answer: In an extraction turbine, steam is withdrawn from one or more stages, at one or more pressures, for heating, plant process, or feed water heater needs. They are often called "bleeder turbines."

131. What is combined-cycle cogeneration?

Answer: A combined cycle using a gas turbine or diesel, usually driving a generator in which the exhaust gases are directed to a waste heat-recovery boiler or heat-recovery steam generator

(HRSG). The steam from the HRSG is then directed to a steam turbo-generator for additional electric power production. The use of the exhaust heat from a gas turbine improves the overall thermal efficiency. In cogeneration, electric power is produced, but part of the steam from the HRSG or from extraction from the steam turbine is used for process heat, hence the term cogeneration-the simultaneous production of electric power and process heat steam.

132. What is important to remember about radial bearings?

Answer: A turbine rotor is supported by two radial bearings, one on each end of the steam cylinder.

These bearings must be accurately aligned to maintain the close clearances between the shaft and the shaft seals, and between the rotor and the casing. If excessive bearing wear lowers the rotor, great harm can be done to the turbine.

133. What is the cause of axial-bore cracks?

Answer: Inadequate toughness of rotor steel and transient thermal stresses.

134. What is the cause of turbine deposits?

Answer: The turbine deposits are steam-born foreign matters settled on turbine blades. Substances dissolved in the BFW transfer partly from the water to steam, during the process of evaporation. They get dissolved in the steam and are carried into the steam turbine.

135. What is the definition of a steam turbine?

Answer: A steam turbine is a prime mover that derives its energy of rotation due to conversion of the heat energy of steam into kinetic energy as it expands through a series of nozzles mounted on the casing or produced by the fixed blades.

1. **Neilson definition**: The turbine is a machine in which a rotary motion is obtained by the gradual change of the momentum of the fluid.
2. **Graham's definition**: The turbine is a prime mover in which a rotary motion is obtained by the centrifugal force brought into action by changing the direction of a jet of a fluid (steam) escaping from the nozzle at high velocity.

136. What is the harm if the rotor is over speed?

Answer: Over speed rotor grows radially causing heavy rub in the casing and the seal system. As a result, considerable amount of shroud-band and tenons-rivet head damage occurs.

137. What is the nature of rotor surface cracks in steam turbines?

Answer: They are shallow in depth and have been located in heat grooves and other small radii at labyrinth-seal areas along the rotor.

138. What is the remedy for a bent steam turbine shaft causing excessive vibration?

Answer:
1. The run-out of the shaft near the centre as well as the shaft extension should be checked.

2. If the run-out is excessive, the shaft is to be replaced.

139. What is the remedy of the damage to blade profiles?
Answer: Upgrading the turbine and depending on the extent of damage, upgrading may involve:
1. Weld repair of affected zones of the blade,
2. Replacement of damaged blades by new ones and of new design,
3. Replacement of base material,
4. Application of protective coatings to guard against corrosion and erosion damage.

140. What is the solution to the problem of SCC/corrosion fatigue of steam turbine blades?
Answer: It involves changing the blade material as well as minimizing the presence of corrodents in steam to a permissible level.

141. What maybe the possible causes for the safety trip to trip at normal speed?
Answer:
1. Excessive vibration.
2. Leakage in the pilot valve.
3. Deposition of dirt in the safety trip valve.

142. What other parts of the steam turbine blades suffer from damage?
Answer:
1. Blade roots.
2. Shroud band.

All Subjects: Mechanical Engineering

01. State all the laws of Thermodynamics?
Answer: Generally thermodynamics contains four laws;
- (A) **Zeroth law:** deals with thermal equilibrium and establishes a concept of temperature.
- (B) **First law:** throws light on concept of internal energy.
- (C) **Second law:** indicates the limit of converting heat into work and introduces the principle of increase of entropy.
- (D) **Third law:** defines the absolute zero of entropy.

02. Explain Second Law of Thermodynamics?
Answer: The entropy of the universe increases over time and moves towards a maximum value.

03. What are the names given to constant temperature, constant pressure, constant volume, constant internal energy, constant enthalpy, and constant entropy processes?
Answer: Isothermal, isochoric, isobaric, free expression, throttling and adiabatic processes respectively.

04. Compare Brayton Cycle and Otto Cycle?
Answer:
- (A) The heat addition and rejection processes in Otto cycle are of constant volume, whereas in Brayton cycle, they are of constant pressure.
- (B) Otto cycle is the ideal cycle for spark ignition engines and Brayton cycle is the ideal cycle for gas power turbines.

05. What is the importance of Thermodynamics?
Answer: All the mechanical engineering systems are studied with the help of thermodynamics. Hence it is very important for the mechanical engineers.

06. How to Measure Temperature in Wet Bulb Thermometer?
Answer: Wet bulb temperature is measured in a wet bulb thermometer by covering the bulb with a wick and wetting it with water. It corresponds to the dew point temperature and relative humidity.

07. Why entropy change for a reversible adiabatic process is zero?
Answer: Because there is no heat transfer in this process.

08. What is the difference between scavenging and supercharging?
Answer: Scavenging is process of flushing out burnt gases from engine cylinder by introducing fresh air in the cylinder before exhaust stroke ends. Supercharging is the process of supplying higher mass of air by compressing the atmospheric air.

09. In a Rankine cycle if maximum steam pressure is increased keeping steam temperature and condenser pressure same, what will happen to dryness fraction of steam after expansion?
Answer: It will decrease.

10. What is the purpose of Scrapper Ring?
Answer: scraps the excess lube oil from the cylinder walls. There by preventing oil from entering combustion zone.

11. What are two essential conditions of perfect gas?
Answer: It satisfies equation of state and its specific heats are constant.

12. Enthalpy and entropy are functions of one single parameter. Which is that?
Answer: Temperature.

13. What is DTSI Technology?
Answer: DTSI stands for Digital Twin Spark Plug Ignition. The vehicles with DTSI Technology use 2 spark plugs which are controlled by digital circuit. It results in efficient combustion of air fuel mixture.
Digital - Since the spark generation will be initiated by a microchip.
Twin - Since two spark plugs will be used.
Spark ignition - Since the ignition will be done via a spark.

14. Why rate of condensation is higher on a polished surface compared to rusty surface?
Answer: Polished surface promotes drop wise condensation and does not wet the surface.

15. How much resistance is offered to heat flow by drop wise condensation?
Answer: No resistance offered.

16. Why Entropy decreases with increase in temperature?
Answer: $ds = dQ/T$, i.e. Entropy is inversely proportional to the temperature. So, as temperature Increases, entropy decreases.

17. What is the relationship between COP of heating and cooling?
Answer: COP of heating is one (unity) more than COP of cooling.

18. How much is the work done in isochoric process?
Answer: Zero.

19. What are the Advantages and Disadvantages of using LPG in Car?
Answer:
 Advantages
1. Complete combustion
2. Fuel saving
3. Homogenous combustion

Disadvantages

1. As complete combustion is occurring, more heat liberated, not advised for long journey, engine will be over heated
2. Installation is difficult
3. Reduce engine life efficiency

20. What are considerations taken into account while creating a piston head?

Answer: The piston head is designed on the basis of the following considerations:

1. The crown should have enough strength to absorb the explosion pressure inside the engine cylinder.
2. The head must always dissipate the heat of the explosion as quickly as possible to the engine walls. The thickness of the head is calculated on the basis of another formula which takes into consideration the heat flowing through the head, the conductivity factor of the material. The temperature at the center and edges of the head.
3. The thickness of the piston head is calculated on the basis of the Grashoff's formula which takes into consideration the maximum gas pressure of an explosion, the permissible bending and the outside diameter of the piston.

21. Why different types of sound are produced in different bikes, though they run on SI Engines?

Answer: Engine specifications are different in different manufactures like as Bore Diameter (CC), Ignition timing. Also the exhaust passage takes more responsible for sound.

22. What should be done to prevent a safety valve to stick to its seat?

Answer: Safety valve should be blown off periodically so that no corrosion can take place on valve and valve seat.

23. Define Octane Number and Cetane Number.

Answer: Octane No: - Octane number is defined as the percentage, by volume, of iso octane in the mixture of iso octane and h-heptane. It is the measure of rating of SI engine.

Cetane No: - Cetane number is defined as the percentage, by volume, of n-cetane in the mixture of n-cetane and alpha methyl naphthalene. It is the measure of rating of CI engine.

24. Why efficiency of gas turbines is lower compared to I.C. engines?

Answer: In gas turbines, 70% of the output of gas turbine is consumed by compressor. I.C. engines have much lower auxiliary consumption. Further combustion temperature of I.C. engines is much higher compared to gas turbine.

25. What do you understand by timed cylinder lubrication?

Answer: For effective lubrication, lube oil needs to be injected between two piston rings when piston is at bottom of stroke so that piston rides in oil during upward movement. This way lot of lube oil can be saved and used properly.

26. What is HUCR in relation to petrol engine?

Answer: HUCR is highest useful compression ratio at which the fuel can be used in a specific test engine, under specified operating conditions, without knocking.

27. Which Mechanism is used in Automobile gearing System?

Answer: Differential mechanism

28. In some engines glycerine is used in place of water for cooling of engine. Why?

Answer: Glycerine has boiling point of 90°C which increases its heat carrying capacity. Thus weight of coolant gets reduced and smaller radiator can be used.

29. Why consumption of lubricating oil is more in two-stroke cycle petrol engine than four-stroke cycle petrol engine?

Answer: In two-stroke engine lube oil is mixed with petrol and thus some lube oil is blown out through the exhaust valves by scavenging and charging air. There is no such wastage in four stroke petrol engine.

30. When Crude Oil is Heated, Which Hydro Carbon comes first?

Answer: Natural gas (Gasoline), at 20 Celsius.

31. What happens if gasoline is used in a Diesel engine? Will diesel engine work?

Answer: No, it will not work, as the Compression ratio of Petrol engine is 6 to 10 & that of Diesel engine is 15 to 22. Thus on such high compression, gasoline gets highly compressed & it may blast.

32. Why boiler is purged every time before starting firing of fuel?

Answer: Purging ensures that any unburnt fuel in furnace is removed; otherwise it may lead to explosion.

33. What is the principle of mechanical refrigeration?

Answer: A volatile liquid will boil under the proper conditions and in so doing will absorb heat from surrounding objects.

34. Why high latent heat of vaporisation is desirable in a refrigerant?

Answer: A high latent heat of vaporisation of refrigerant results in small amount of refrigerant and thus lesser circulation system of refrigerant for same tonnage.

35. As compression ratio increases, thermal n increases. How is thermal n affected by weak and rich mixture strength?

Answer: Thermal n is high for weak mixture and it decreases as mixture strength becomes rich.

36. How a diesel engine works in generator?

Answer: Diesel engine is a prime mover, for a generator, pump, and for vehicles etc. Generator is connected to engine by shaft. Mostly in thermal power plant, there is an engine is used to drive generator to generate power.

37. What is flashpoint?
Answer: Flash point: the lowest temperature at which the vapour of a combustible liquid can be ignited in air.

38. What is the critical temperature of a refrigerant?
Answer: Critical temperature is the maximum temperature of a refrigerant at which it can be condensed into liquid and beyond this it remains gas irrespective of pressure applied.

39. What is refrigerant?
Answer: Any substance that transfers heat from one place to another, creating a cooling effect. Water is the refrigerant in absorption machines.

40. What is the effect of reheat on Rankine cycle?
Answer: This prevents the vapour from condensing during its expansion which can seriously damage the turbine blades, and improves the efficiency of the cycle, as more of the heat flow into the cycle occurs at higher temperature.

41. How engine design needs to be changed to burn lean mixture?
Answer: Engine to burn lean mixture uses high compression ratio and the highly turbulent movement of the charge is produced by the geometry of the combustion chamber.

42. Horse power of I.C. engines can be expressed as RAC rating, SAE rating, or DIN rating. To which countries these standards belong?
Answer: U.K., USA and Germany respectively.

43. Why are Head Gaskets blown?
Answer: Normally head gasket blows, when the engine overheats and they can also blow from incorrect installation or poor design. Head gaskets expand and contract according to engine temperature, these cycles may happen after a long period of time, causes the gasket to fail. If you're replacing the gasket, check the engine block, and head for warping. Follow proper cleaning and torque specifications during assembly.

44. What is the Difference between a Humidifier and Vaporizer?
Answer:

(A) The basic difference between humidifiers and vaporizers is that humidifiers disperse cool mist into the air, and vaporizers heat the water to disperse hot steam. Humidifiers are normally used in cooler climates, when due to the usage of heater, the air in the house becomes too dry for comfort and also, it becomes very difficult to breathe. Humidifiers release cool moisture droplets into the air.

(B) Vaporizers also help in moistening the dry air in the house, but vaporizers release hot vapour into the air. There is a heating element in the vaporizers, which help in releasing steam.

(C) Vaporizers heat the water and then release vapours.

(D) The basic difference between them is that one emits cold vapours, while the other one emits hot vapours.

45. What do you understand by fuel cycle in nuclear plants?
Answer: Fuel cycle a series of sequential steps involved in supplying fuel to a nuclear power reactor. The steps include : Mining, refining uranium, fabrication of fuel elements, their use in nuclear reactor, chemical processing to recover remaining fissionable material, re-enrichment of fuel from recovered material, re-fabrication of new fuel elements, waste storage etc.

46. What is heavy water and what is its use in nuclear plants?
Answer: Water containing heavy isotopes of hydrogen (Deuterium) is known as heavy water. Heavy water is used as a moderator. Heavy water has low cross section for absorption of neutrons than ordinary water. Heavy water slows down the fast neutrons and thus moderates the chain reaction.

47. What is a converter reactor?
Answer: A reactor plant which is designed to produce more fuel than it consumes. The breeding is obtained by converting fertile material to fissile material.

48. Difference between Absorption and Adsorption
Answer: In absorption, one substance (matter or energy) is taken into another substance. But in adsorption only the surface level interactions are taking place.

49. Is the boiler a closed system?
Answer: Yes definitely the boiler is a closed system.

50. Difference between Gas and Vapour
Answer:
(A) Vapour can turn back and forth into liquid and solid states but a gas cannot.
(B) Gases cannot be seeing while vapours are visible.
(C) Vapours settle down on ground while gases do not.

51. Difference between Boiling Point and Melting Point
Answer:
(A) The melting point is a defined for solids when it transfers from solid state to liquid state.
(B) The boiling point is defined for liquids for a state change from liquid to gas.
(C) Boiling point is highly dependent on the external pressure whereas the melting point is independent of the external pressure.

52. State Laws of conservation of energy?
Answer: According to the laws of conservation of energy, "energy can neither be created nor be destroyed. It can only be transformed from one form to another."

53. Difference between Liquid and Aqueous

Answer: Liquid is a state of matter, while aqueous is a special type of liquid formed by dissolving a compound in water

All aqueous solutions are liquids, but not all liquids are aqueous solutions.

54. What is Carnot engine?

Answer: It was being designed by Carnot and let me tell you that Carnot engine is an imaginary engine which follows the Carnot cycle and provides 100% efficiency.

55. What is the use of flash chamber in a vapour compression refrigeration cycle to improve the COP of refrigeration cycle?

Answer: When liquid refrigerant as obtained from condenser is throttled, there are some vapours. These vapours if carried through the evaporator will not contribute to refrigerating effect. Using a flash chamber at some intermediate pressure, the flash vapour at this pressure can be bled off and fed back to the compression process. The throttling process is then carried out in stages. Similarly compression process is also done in two separate compressor stages.

56. Why pistons are usually dished at top?

Answer: Pistons are usually hollowed at top to
  (A) Provide greater space for combustion
  (B) Increase surface for flue gases to act upon, and
  (C) Better distribution of stresses.

57. What is the function of thermostat in cooling system of an engine?

Answer: Thermostat ensures optimum cooling because excessive cooling decreases the overall efficiency. It allows cooling water to go to radiator beyond a predetermined temperature.

58. Which formula forms a link between the Thermodynamics and Electro chemistry?

Answer: Gibbs Helmholtz formula is the formula which forms the link between the thermodynamics and electromagnetism.

$$\Delta Hs/R = [\partial \ln.p /\partial(1/T)] (x)$$

Where: x – mole fraction of CO_2 in the liquid phase

  p – CO_2 partial pressure (kPa)
  T – Temperature (K)
  R – Universal gas constant
  α – mole ratio in the liquid phase (mole CO_2 per mole of amine)

59. What is Hess Law?

Answer: According to the Hess law the energy transfer is simply independent of the path being followed.

If the reactant and the product of the whole process are the same then same amount of energy will be dissipated or absorbed.

60. Which has more efficiency: Diesel engine or Petrol engines?

Answer: Off course Diesel engine has the better efficiency out of two.

61. Why iso-octane is chosen as reference fuel for S.I. engines and allotted 100 values for its octane number?
Answer: Iso-octane permits highest compression without causing knocking.

62. How is natural gas better than LPG?
Answer: Without getting into the chemistry and physics of the two different types of gases, natural gas has a higher but output than liquid propane gas. In other words, higher the available energy output/more energy is efficient.

63. Why thermal efficiency of I.C. engines is more than that of gas turbine plant?
Answer: In I.C. engine maximum temperature attained is higher than in gas turbine.

64. Which are the reference fuels for knock rating of S.I. engines?
Answer: n-heptane and ISO-octane.

65. What will be the position of Piston Ring?
Answer: In 180 degree angle the Top ring, Second ring and Oil ring are fixed. Position the ring approximately 1 inch gap below the neck.

66. What is Heat rate of a Power plant?
Answer: Heat rate is a measure of the turbine efficiency. It is determined from the total energy input supplied to the Turbine divided by the electrical energy output.

67. When effect of variations in specific heats is considered then how do maximum temperature and pressure vary compared to air standard cycle?
Answer: Temperature increases and pressure decreases.

68. Quantities like pressure, temperature, density, viscosity, etc. are independent of mass. What are these called?
Answer: Intensive properties.

69. Why we do not use same technology to start both SI /CI engine?
Answer: The S.I. or spark ignition engine uses petrol as a fuel and the C.I. or compression ignition engine uses diesel as a fuel. Both the fuels has different compression ratio. In SI engine the compression ratio is 8-12:1. In CI engine the compression ratio is 16-22:1. So in case of SI engine, the compression ratio is not sufficient for fuel to burn so a spark plug is used, whereas in CI engine, the compression ratio is so high that due to its internal heat the fuel is combusted so there is no need for a spark plug. So the technology used in SI engine is different from CI engine.

70. What VVTi written on new cars of Toyota stands for?
Answer: VVTi: Variable Valve Timing with Intelligence. It is the advanced version of the VVT engine.

It changes the cam shaft position by using oil pressure. It is similar to CVVT in Hyundai.

71. In convection heat transfer, if heat flux intensity is doubled then temperature difference between solid surface and fluid will?
Answer: Get doubled.

72. How you can define coal?
Answer: Coal is a naturally occurring hydrocarbon that consists of the fossilised remains of buried plant debris that have undergone progressive physical and chemical alteration, called coalification, in the course of geologic time.

73. Why is the Suction pipe of Vapour Compression Refrigeration system insulated?
Answer:
   (A) It prevents the suction line from sweating and dripping water inside the house.
   (B) The insulation also prevents the suction line attracting heat from the outdoors on its way to the condenser coil.

74. How to determine the capacity of Refrigeration system? How we use Condenser coils, Compressor, Capillarity?
Answer: To determine the Refrigeration system by test of C.O.P and use the Condenser coils, Compressor, Capillarity, based on the properties

75. What do you mean by super critical above 500MW in Thermal power plant?
Answer: In super critical boiler means all the steam to convert in to superheated steam at outlet of boiler no need to sent to super heater once again.

76. Which pollutant is major greenhouse gas and what is its effect?
Answer: CO is major greenhouse gas and it traps the radiation of heat from the sun within earth's atmosphere.

77. In order to increase efficiency and reduce CO emissions and other emissions, clear coal technologies are receiving major attention. What are these?
Answer:
   (A) Advanced pulverised and pressurised pulverised fuel combustion.
   (B) Atmospheric fluidised bed combustion and pressurised fluidised bed combustion.
   (C) Supercritical boilers.
   (D) Integrated gasification combined cycle systems.
   (E) Advanced integrated gasification, including fuel cell systems.
   (F) Magneto hydrodynamic electricity generation.

78. What does CC Stand for?
Answer: CC is the abbreviated form of cubic centimetre. It is the unit by which the capacity of an engine is designated. It is the volume between TDC and BDC. It represents the quantity of fuel-air mix or exhaust gas that is pumped out in a single piston stroke. Alternatively it can represent the volume of the cylinder itself.

79. State the difference between ultimate and proximate analysis of coal?

Answer: In ultimate analysis, chemical determination of following elements is made by weight: Fixed and combined carbon, H, O, N, S, water and ash. Heating value is due to C, H and S. In proximate analysis following constituents are mechanically determined by weight. Moisture, volatile matter, fixed carbon and ash heating value is due to fixed carbon and volatile matter.

80. What is fuel ratio?

Answer: Fuel ratio is the ratio of its % age of fixed carbon to volatile matter.

81. How the analyses and calorific values of fuels can be reported?

Answer: It may be reported as:
- (A) As received or fired (wet) basis
- (B) Dry or moisture free basis
- (C) Combustible or ash and moisture free basis

82. We have read that when the piston goes up and down then the engine works i.e. the suction, compression etc. then what happens in the case of big vehicles, which start at stable condition, i.e. how does their piston moves when they are at rest. How suction, compression etc

Answer: Smaller vehicles like bikes, cars are started with the help of motors. Initially, motors turn the crank shaft till sufficient suction pressure is reached. When sufficient suction pressure is reached, the engine starts to suck the fuel in and then the cycle begins when the fuel is taken in and ignited. Similarly, for huge engines, instead of motors, we use starting air. Air at a pressure of 10-30 bar is fed to the engine which is at rest. This air rotates the engine till it attains sufficient suction pressure. Once the pressure is reached, the cycle starts and it starts firing.

83. The Compression ratio of Petrol engine is always less than Compression Ratio of Diesel engine why?

Answer: Petrol is not self igniting; it needs spark to flame up in chamber. Whereas diesel is self igniting in diesel engine, to attain that state it requires high temp &pressure. This temperature & pressure is more than what's required in Petrol Engines by property of that fluid.

84. What are the important operational performance parameters in design of fuel firing equipment?

Answer: Fuel flexibility, electrical load following capability, reliability, availability, and maintenance ease.

85. What is the difference between total moisture and inherent moisture in coal?

Answer: The moisture content of the bulk as sampled is referred to as total moisture, and that of the air dried sample is called inherent moisture.

86. Ultimate analysis of coal is elementary analysis. What it is concerned with?

Answer: Carbon, hydrogen, nitrogen, and sulphur in coal on a weight percentage basis.

87. What is the temperature of space?

Answer: The short answer is that the temperature in space is approximately 2.725 Kelvin. That means the universe is generally just shy of three degrees above absolute zero, the temperature at which molecules themselves stop moving. That's almost -270 degrees Celsius, or -455 Fahrenheit.

88. Proximity analysis of coal provides data for a first, general assessment of a coal's quality and type. What elements it reports?
Answer: Moisture, volatile matter, ash and fixed carbon.

89. What are the causes of main engine black smoke?
Answer: There is many cause of black smoke.
   (A) Is improper mixture of fuel supply by carburettor like very rich mixture so the fuel improper burn.
   (B) It is when piston or piston ring is fail so back side cooling oil release in combustion chamber it cause black smoke.
   (C) Improper ignition system like not sufficient time of pressure rise delay period.

90. What is the significance of torque (in Nm) given in the engine specification?
Answer: it give the moment about any point or simple rotation.

91. Explain the difference between AFBC, BFBC, PFBC and CFBC in regard to fluidised bed technologies.
Answer:
AFBC (Atmospheric fluidised bed combustion) process consists of forming a bed of inert materials like finely sized ash or ash mixed with sand, limestone (for sulphur removal), and solid fuel particles in a combustor and fluidising it by forcing combustion air up through the bed mixture. The gas flows through bed without disturbing particles significantly but gas velocity is high enough to support the total weight of bed (fluidisation). At slightly higher velocity excess gas passes through the bed as bubbles (fluidised bed) and gives the bed the appearance of a boiling liquid.

BFBC (Bubbling fluidised bed combustion) has a defined height of bed material and operates at or near atmospheric pressure in the furnace.

PFBC (Pressurised fluidised bed combustion) system operates the bed at elevated pressure. Exhaust gases have sufficient energy to power a gas turbine, of course, gases need to be cleaned. In fluidised combustion, as ash is removed some unburned carbon is also removed resulting in lower efficiency.

CFBC (circulating fluidised bed combustion) system, bed is operated at higher pressure leading to high heat transfer, higher combustion efficiency, and better fuel feed. Circulating fluidised beds operate with relatively high gas velocities and fine particle sizes. The maintenance of steady state conditions in a fast fluidised bed requires the continuous recycle of particles removed by the gas stream (circulating bed). The term circulating bed is often used to include fluidised bed systems containing multiple conventional bubbling beds between which bed material is exchanged.

92. What is the difference between nuclear fission and fission chain reaction?

Answer: The process of splitting of nucleus into two almost equal fragments accompanied by release of heat is nuclear fission. Self sustained, continuing, sequence of fission reactions in a controlled manner is fission chain reaction.

93. What is BHP?

Answer: Brake horsepower is the amount of work generated by a motor under ideal conditions. This work is calculated without the consideration of effects of any auxiliary component that may slow down the actual speed of the motor. Brake horsepower is measured within the engines output shaft and was originally designed to calculate and compare the output of steam engines. As per the conventions, 1 BHP equals to:
- (1) 745.5 watts
- (2) 1.01389 ps
- (3) 33,000 ft lbf/min
- (4) 42.2 BTU/min

94. Which parameter remains constant in a throttling process?
Answer: Enthalpy.

95. What is the difference between isentropic process and throttling process?
Answer: In isentropic process, heat transfer takes place and in throttling process, enthalpy before and after the process is same.

96. What is D4D Technology used in Toyota Vigo?

Answer: D-4D is widely recognized as one of the most advanced diesel technologies on the market today. Diesel engines relied on relatively simple technology with a low-pressure mechanical injector delivering fuel to a pre-combustion chamber in the cylinder head where a single ignition – fired by the intense heat of high compression – takes place.

97. What for Schmidt plot for is used in heat transfer problems?
Answer: Schmidt plot is a graphical method for determining the temperature at any point in a body at a specified time during the transient heating or cooling period.

98. What is ATFT Technology used in Honda Hunk?

Answer: ATFT means Advance Tumble Flow Induction Technology, Tumble flow means swirling. In this technology, fuel air mixture from the carburettor into the engine cylinder with a swirl action. The advantage being one gets a more efficient burning of fuel hence more power and better fuel economy with lesser emissions.

99. What is big advantage of fast breeder reactor?
Answer: It has rapid self breeding of fissile fuel during the operation of the reactor, and thus, it offers about sixty times the output with same natural uranium resources through ordinary non breeder nuclear reactor.

100. What is the purpose of biological shield in nuclear plants?

Answer: Biological shield of heavy concrete prevents exposure to neutrons, beta rays and gamma rays which kill living things.

101. When a real gas behaves like ideal gas?
Answer: A real gas behaves like an ideal gas in low pressure and high temperature conditions.

102. At which temperature thermal radiation can become zero?
Answer: Not possible. Because thermal radiation becomes only zero at absolute zero temperature which can never be attained by the third law of thermodynamics.

103. What is the significance of entropy?
Answer: As per the second law of thermodynamics, any heat input to the system (Heat engine) cannot be converted completely into useful work. Some energy is lost and that is called 'unavailable work'. The amount of unavailable work increases as the entropy increases.

104. What is the difference between heat transfer and thermodynamics?
Answer: Heat transfer deals with the energy analysis which in transition and depends on the modes of heat transfer like conduction, convection and radiation or combination of any modes. Heat transfer deals in non equilibrium domain and conditions while thermodynamics deals with study of system at equilibrium and does not depend on how heat transfer is calculated.

105. How catalyst converter works?
Answer: In Fuel Cell, a catalyst is a substance that causes or accelerates a chemical reaction without itself being affected. Catalysts participate in the reactions, but are neither reactants nor products of the reaction they catalyze.

106. What will happen if relief valve in hydraulic system fails?
Answer: The main function of pressure relief valve is to maintain the pressure in hydraulic system. It is one mounting which is used for safety. When pressure increases then safety valve comes into action & if the valve get fail the system get damage due to excessive pressure.

107. How can problems of "excessive vibration or noise" due to piping strain be avoided on steam turbines?
Answer:
1. The inlet as well as exhaust steam lines should be firmly supported to avoid strains from being imposed on the turbine.
2. Adequate allowance should be made for expansion of steam pipes due to heat.

108. How the deposits in turbine be removed?
Answer: Deposits in turbine can be removed by
1. Water soluble deposits may be washed off with condensate or wet steam.
2. Water insoluble deposits are removed mechanically after dismantling the turbine.
Experience shows that water soluble deposits are embedded in layers of water insoluble deposits. And when the washing process is carried out, water soluble parts of the deposit

dissolve away leaving a loose, friable skeleton of water-insoluble deposits which then break loose and wash away.

109. How the fatigue damage on high-pressure blades be corrected?

Answer: Fatigue-damage on high-pressure blades arises due to vibration induced by partial arc admission. This can be corrected by switching over to full arc admission technique.

110. How the misalignment of Flanges be rectified?

Answer: The bolts holding the flanges together are to be tightened. The coupling is to be checked for sureness between the bore and the face. At the same time axial clearance is to be checked.

111. How the problem of excessive speed variation due to throttle assembly friction be overcome?

Answer: The throttle should be dismantled. Moving parts should be checked for free and smooth movement. Using very fine-grained emery paper, the throttle valve seats and valve steam should be polished.

112. How the problems of vibration and fatigue arise in steam turbine blades?

Answer:
1. These arise due to flow irregularities introduced because of manufacturing defects, e.g. lack of control over tolerances.
2. System operating parameter, e.g. low flow may excite various modes of vibration in the blades.

113. How does solid-particle erosion occur?

Answer: Solid-particle erosion, i.e. SPE occurs in the high-pressure blades. And it takes place when hard particles of iron exfoliated by steam from super-heater tubes, reheater tubes, steam headers and steam leads strike on the surface of turbine blades.

114. How does the internal efficiency monitoring lead to the detection of turbine deposits?

Answer:
1. Process heat drop.
2. Adiabatic heat drop.
3. The process heat drop and adiabatic heat drop are obtained from a Moliere- Chart for the corresponding values of steam parameters – pressure and temperature – at initial and final conditions.

115. How is a flyball governor used with a hydraulic control?

Answer: As the turbine speeds up, the weights are moved outward by centrifugal force, causing linkage to open a pilot valve that admits and releases oil on either side of a piston or on one side of a spring-loaded piston. The movement of the piston controls the steam valves.

116. How does axial thrust balance in multistage pump?

Answer: A balancing line from discharge end is connected to suction side to balance axial thrust.

117. How to calculate the boiler efficiency? Any formula is there?

Answer: boiler efficiency= (heat transferred to feed water in converting it to steam)/ (heat released by complete combustion of fuel)

η (eta) = Mass of steam × [h - H(water)]/(mass of fuel × calorific value fuel)

118. How cooling tower height selected?

Answer: The Function of a cooling tower is to cool the water coming from condenser. The water coming from condenser is hot and it is sprayed in a cooling tower and a air coming out from bottom cool the water which is coming down. Outside air is cool and air in inside the cooling tower is hot due to humidity. So there is a density difference between outside and inside air which caused pressure difference.

Pr Difference = g × H × (density difference)

Where,

H = Height of chimney

Pr Difference = Pr Difference so that air can flow to cooling tower from outside

119. What is the exact requirement of priming?

Answer: priming is done in pumps to remove the entrapped air from the suction pipe thus aiding in smooth operation and avoiding in excess load on the pump.

120. How tonnage can be controlled in PLC base hydraulic press

Answer: custom integration of press interlocks to inter facing with other parts of the hydraulic press line such as the feeder or transfer systems.

121. Besides lubrication, what are two functions of lubricating oil in some turbines?

Answer:

In larger units, lube oil cools the bearings by carrying off heat to the oil coolers.

Lube oil in some turbines also acts as a hydraulic fluid to operate the governor speed-control system.

122. Which two elements have same percentage in proximate and ultimate analysis of coal?

Answer: Moisture and ash.

123. On which analysis is based the Delong's formula for the heating value of fuel?

Answer: On ultimate analysis.

124. Explain nuclear reactor in brief.

Answer: A plant which initiates sustains controls and maintains nuclear fission chain reaction and provides shielding against radioactive radiation is nuclear reactor.

125. What is the difference between conversion and enrichment?

Answer: The process of converting the non fissile U 38 to fissile U35 is also called "Conversion". The material like U 38 which can be converted to a fissile material by the neutron flux is called "fertile material". The conversion is obtained within the nuclear reactor during the chain reaction. Enrichment is the process by which the proportion of fissile uranium isotope (U35) is increased above 0.7% (original % in natural uranium). The concentration of U35 in the uranium hexafluoride is increased from the 0.7% in natural uranium to 4%. This is called enrichment and is accomplished in an enrichment plant.

126. Which element causes difference in higher and lower heating values of fuel?
Answer: Hydrogen.

127. What is PGM FI technology used in Honda Stunner Bike?
Answer: The development of an ECU-integrated throttle body module for an electronic fuel injection system for small motorcycles. Honda has a goal to reduce the total emissions of HC (hydro-carbon) from new vehicles to approximately 1/3 and to further improve the average fuel economy by approximately 30 (both from 1995) by the year 2005. To realize the goal we at Asaka R&D Center considered that the small motorcycles used in many countries in the world should be improved further for clean exhaust gas and low fuel consumption. Accordingly we have started development of the PGM-FI system for small motorcycles with engines of 125cc or smaller including air-cooled engines. To ensure clean exhaust gas and high fuel economy the control of combustion through an accurate fuel supply is a must. As the conventional FI system (electronic fuel injection system) applied to motorcycles is bulky and costly its application has been mostly in large motorcycles using multi-cylinder engines. In the newly developed PGM-FI in order to apply to small displacement models the obstacles have been eliminated by fully using Honda's techniques to down-size components as well as making maximum use of the FI techniques attained from the large motorcycles. The compact PGM-FI offers new benefits such as the reduction of released environmentally detrimental substances and the improvement of drivability economy etc

128. Disposal of radioactive waste materials and spent fuel is a major and important technology. How the waste radioactive material is disposed off?
Answer: Non usable fission products are radioactive and take short/medium/long time for radioactive decay to reach safe level of radioactivity.
Accordingly three methods of disposal are:
   (A) Zero or low radioactivity material is dispersed or stored without elaborate shielding.
   (B) Medium radioactivity material is stored for short duration of about 5 years to allow decay of radioactivity.
   (C) High radioactive material. They are stored in water for several months to permit radioactive decay to an acceptable low level.

129. Which nuclear reactor uses water as a coolant, moderator and reflector?
Answer: Pressurised water reactor.

130. Which reactor produces more fissionable material than it consumes?
Answer: Breeder reactor.

131. Why water can't be used as refrigerant for small refrigerating equipment?
Answer: The refrigerant should be such that vapour volume is low so that pumping work will be low. Water vapour volume is around 4000 times compared to R-for a given mass.

132. Explain the second law of thermodynamics.
Answer: The entropy of the universe increases over time and moves towards a maximum value.

133. What kinds of pipes are used for steam lines?

Answer: Normally galvanized pipes are not used for steam. Mild steel with screwed or welded fittings are the norm. Pressure and temperature are very important factors to be considered in what type of materials to be used. Steam even at low pressures can be extremely dangerous.

134. What is the difference between shear center flexural center of twist and elastic center?

Answer: The shear center is the centroid of a cross-section. The flexural center is the center of twist, which is the point on a beam that you can add a load without torsion. The elastic center is located at the center of gravity. If the object is homogeneous and symmetrical in both directions of the cross-section then they are all equivalent.

135. What is the difference between projectile motion and a rocket motion?

Answer: A projectile has no motor/rocket on it, so all of its momentum is given to it as it is launched. An example of a projectile would be pen that you throw across a room.
A rocket or missile does have a motor/rocket on it so it can accelerate itself while moving and so resist other forces such as gravity.

136. What is ferrite?

Answer: Magnetic iron rock

137. What is a cotter joint?

Answer: These types of joints are used to connect two rods, which are under compressive or tensile stress. The ends of the rods are in the manner of a socket and shaft that fit together and the cotter is driven into a slot that is common to both pieces drawing them tightly together. The tensile strength of the steel is proportionate to the strength needed to offset the stress on the material divided by the number of joints employed.

138. What is the alloy of tin and lead?

Answer: A tin and lead alloy is commonly called solder. Usually solder is a wire with a rosin core used for soldering. The rosin core acts as a flux.

139. What does F.O.F. stand for in piping design?

Answer: FOF stands for Face of Flange. A flange has either of the two types of faces:
  a) Raised face
  b) Flat face
The F.O.F is used to know the accurate dimension of the flange in order to avoid the minute errors in measurement in case of vertical or horizontal pipelines.

140. Explain Otto cycle.

Answer: Otto cycle can be explained by a pressure volume relationship diagram. It shows the functioning cycle of a four stroke engine. The cycle starts with an intake stroke, closing the intake and moving to the compression stroke, starting of combustion, power stroke, heat exchange stroke where heat is rejected and the exhaust stroke. It was designed by Nicolas Otto, a German engineer.

141. What is gear ratio?
Answer: It is the ratio of the number of revolutions of the pinion gear to one revolution of the idler gear.

142. What is annealing?
Answer: It is a process of heating a material above the re-crystallization temperature and cooling after a specific time interval. This increases the hardness and strength if the material.

143. What is enthalpy?
Answer: Enthalpy is the heat content of a chemical system.

144. What is ductile-brittle transition temperature?
Answer: It is the temperature below which the tendency of a material to fracture increases rather than forming. Below this temperature the material loses its ductility. It is also called Nil Ductility Temperature.

145. What is a uniformly distributed load?
Answer: A UDL or uniformly distributed load is a load, which is spread over a beam in such a way that each unit length is loaded to the same extent.

146. What are the differences between pneumatics and hydraulics?
Answer: The differences between pneumatics and hydraulics are:
  a) Working fluid: Pneumatics use air, Hydraulics use Oil
  b) Power: Pneumatic power less than hydraulic power
  c) Size: P components are smaller than H components
  d) Leakage: Leaks in hydraulics cause fluid to be sticking around the components. In pneumatics, air is leaked into the atmosphere.
  e) Pneumatics obtain power from an air compressor while hydraulics require a pump
  f) Air is compressible, hydraulic oil is not

147. What is a positive displacement pump?
Answer: A positive displacement pump causes a liquid or gas to move by trapping a fixed amount of fluid or gas and then forcing (displacing) that trapped volume into the discharge pipe. Positive displacement pumps can be further classified as either rotary-type (for example the rotary vane) or lobe pumps similar to oil pumps used in car engines. These pumps give a non-pulsating output or displacement unlike the reciprocating pumps. Hence, they are called positive displacement pumps.

148. Why would you use hydraulics rather than pneumatics?

Answer: Hydraulics is suitable for higher forces & precise motion than pneumatics. This is because hydraulic systems generally run at significantly higher pressures than pneumatics systems. Movements are more precise (repeatable) because hydraulics uses an incompressible liquid to transfer power whilst pneumatics uses gases. Pneumatic systems have some advantages too. They are usually significantly cheaper than hydraulic systems, can move faster (gas much less viscous than oil) and do not leak oil if they develop a leak.

149. What is isometric drawing?
Answer: It is a 3-D drawing used by draftsmen, architects etc

150. What are the advantages of gear drive?
Answer: In general, gear drive is useful for power transmission between two shafts, which are near to each other (at most at 1m distance). In addition, it has maximum efficiency while transmitting power. It is durable compare to other such as belts chain drives etc. You can change the power to speed ratio.

Advantages: -
   a) It is used to get various speeds in different load conditions.
   b) It increases fuel efficiency.
   c) Increases engine efficiency.
   d) Need less power input when operated manually.

151. Which conducts heat faster steel copper or brass?
Answer: Copper conducts heat faster than steel or brass. Any material that is good for conducting heat is also good for electricity in most cases. Wood terrible for transferring heat thus is also insulator for electric.

152. What is a Process Flow Diagram?
Answer: A Process Flow Diagram (or System Flow Diagram) shows the relationships between the major components in the system. It also has basic information concerning the material balance for the process.

153. How pipe flanges are electrically insulated?
Answer: Pipe flanges are protected from corrosion by means of electrolysis, with dielectric flanges. The piping system is electrically insulated by what is called a sacrificial anode. A bag of readily corrodible metal is buried in the ground with a wire running from the pipe to the bag so that the sacrificial anode will corrode first. If any electrical current charges the pipe, it also serves as a ground.

154. Where pneumatic system is used?
Answer: Any system needs redundancy in work needs pneumatics, because the compressor of the pneumatic system has periodical operations (intermittent work, not as hydraulic pump). The

compressed air could be accumulated in tanks with high pressures and used even if the compressor failed.

155. Why gas containers are mostly cylindrical in shape?
Answer: The most efficient shape for withstanding high pressure is a sphere but that would be costly to manufacture. A cylinder with a domed top and a domed bottom (look underneath, the flat base is actually welded around the outside, the bottom of the gas container is actually domed) is a much cheaper shape to manufacture whilst still having good strength to resist the internal gas pressure.

156. How is martensite structure formed in steel?
Answer: Martensite transformation begins when austenite is cooled below a certain critical temperature, called the matrensite start temperature. As we go below the martensite start temperature, more and more martensite forms and complete transformation occurs only at a temperature called matrensite finish temperature. Formation of martensite requires that the austenite phase must be cooled rapidly.

157. What is an orthographic drawing?
Answer: Orthographic projections are views of a 3D object, showing 3 faces of it. The 3 drawings are aligned so that if the page were folded, it would create part of the shape. It is also called multiview projections.
The 3 faces of an object consist of its plan view, front view and side view. There are 2 types of orthographic projection, which are 1st angle projection and 3rd angle projection.

158. What is representative elementary volume?
Answer: Smallest volume over which measurements can be made that will yield a representative of the whole.

159. Why are LNG pipes curved?
Answer: LNG pipes are curved because LNG is condensed gas (-164° Celsius) so it can expand the pipes that is what engineers designed the LNG pipes are curve type.

160. What does angular momentum mean?
Answer: Angular momentum is an expression of an objects mass and rotational speed. Momentum is the velocity of an object times it is mass, or how fast something is moving how much it weigh. Therefore, angular momentum is the objects mass times the angular velocity where angular velocity is how fast something is rotating expressed in terms like revolutions per minute or radians per second or degrees per second.

161. Can you use motor oil in a hydraulic system?

Answer: Hydraulic fluid has to pass a different set of standards than motor oil. Motor oil has tackifiers, lower sulphur content, and other ingredients that could prove harmful to the seals and other components in a hydraulic system. If it is an emergency only should you do it.

162. What causes white smoke in two stroke locomotive engines?

Answer: That is the engine running too lean (lack of fuel). This condition will lead to overheating and failure of the engine.

163. What is the role of nitrogen in welding?

Answer: Nitrogen is used to prevent porosity in the welding member by preventing oxygen and air from entering the molten metal during the welding process. Other gases are also used for this purpose such as Argon, Helium, Carbon Dioxide, and the gases given off when the flux burns away during SMAW (stick) welding.

164. What does Green field project mean?

Answer: Green field projects are those projects, which do not create any environmental nuisance
(Pollution), follows environmental management system and EIA (environment impact assessment). These projects are usually of big magnitude.

165. Is it the stress that, produces strain or strain produces stress?

Answer: A Force applied to an object will cause a displacement. Strain is effectively a measure of this displacement (change in length divided by original length).
Stress is the Force applied divided by the area it is applied. (E.g. pounds per square inch)
Therefore, to answer the question, the applied force produces both "Stress and Strain". "Stress and Strain" are linked together by various material properties such as Poisson's ratio and Young's Modulus.

166. How does iron ore turn into steel?

Answer: To make Steel, Iron Ore is refined into iron and all the carbon is burned away using very high heat (Bessemer). A percentage of Carbon (and other trace elements) are added back to make steel.

167. What is knurling?

Answer: Knurling is a machining process normally carried out on a centre lathe. The act of Knurling creates a raised criss-cross pattern on a smooth round bar that could be used as a handle or something that requires extra grip.

168. What is the mechanical advantage of a double pulley?

Answer: It only takes half the effort to move an object but twice the distance.

169. What is extrude aluminium?

Answer: Extrusion is the process where a metal or a metal bar is pulled through a mandrel to elongate it and/or give it a final shape.

Extruded Aluminium is a common form of making small aluminium wire, bars or beams and many varieties of small non-structural, decorative pieces.

170. What is a Newtonian fluid?

Answer: A Newtonian fluid possesses a linear stress strain relationship curve and it passes through the origin. The fluid properties of a Newtonian fluid do not change when any force acts upon it.

171. What are the points in the stress strain curve for steel?

Answer: Proportional limit, elastic limit or yield point, ultimate stress and stress at failure.

172. What is a feasibility study?

Answer: In order to make wise investments in a marketplace experiencing increasing levels of risk, companies are turning to feasibility studies to determine if they should offer new products, services or undertake a new business endeavour. The purpose of a feasibility study is to determine if a business opportunity is possible, practical, and viable. When faced with a business opportunity, many optimistic people tend to focus on just the positive aspects. A feasibility study enables a realistic view at both the positive and negative aspects of the opportunity. A feasibility study is an important tool for making the right decisions. A wrong decision often leads to business failure. For example, only 50% of start-ups are still in business after 18 months and only 20% are in business after 5 years.

173. Is it possible for something to explode in space like in the movies?

Answer: Yes, provided there is an oxygen source. A spaceship with liquid hydrogen and liquid oxygen will blow up quite well in the vacuum of space. Chemical explosives will also explode in space since they function by breaking weakly bonded chemical components; no oxygen is necessary. Nuclear explosions can of course occur in space, too. The United States military in the 60's performed a series of nuke tests in outer space, and found out what EMP can do, when they briefly wiped out Hawaii's electrical grid for a few hours.

174. What are the minerals needed to make or build a car?

Answer:-
   a) Sulphur is used in tire construction
   b) Aluminium and iron in the car body and engine
   c) Quartz is used in windshield construction
   d) Chromium is used for chrome plating
   e) Copper is used in wiring
   f) Magnesium may be used in the wheels
   g) Zinc would be used in rust plating
   h) Gold may be used in the spark plugs and in electrical connections

i) Tin may be used in solder
j) Lead will be used in the batteries
k) Tungsten may be used in the light bulb filaments

175. How do you determine what suction pressure should be in R22 air conditioning compressors suction line?

Answer: The type of refrigerant used and sometimes the amount of charge determines the proper suction pressure of any air conditioning or refrigeration system.

176. How do motion detectors work?

Answer: Motion detectors work to help with construction work.

177. If the black box flight recorder is never damaged during a plane crash, why not the whole airplane made out of that stuff?

Answer: The SR-71 was made mostly out of Titanium (very high strength to weight ratio) and it got off the ground fine. In addition, a two-seat airplane cost $33 million in 1960s dollars, so no one would ever be able to afford an airliner made out of it.

178. Does the position of a chicken egg affect the amount of weight it can withstand?

Answer: Yes. Lying on its side, it will break with very little pressure. That is why, when you break an egg, you always strike the side.

179. How do you repair a hydraulic jack?

Answer: In general, it seems simple: polishing scored cylinders and pistons, replacing worn seals, replacing excessively worn or damaged parts, etc. However, for an ordinary person, the problem I have run into is finding a readily accessible source of parts, as the hydraulic repair industry seems to be a closed fraternity.

180. Where do you put the oil in a compressor?

Answer: In the reciprocating compressors, you fill the crankcase with the lubricante.
In screw compressors, there is a reservoir to separate oil from compressed air, that reservoir works as the oil sump of the screw compressor.

181. What is a pneumatic system?

Answer: Pneumatic system is a system that uses air to power something. For instance, have you seen the tube systems at bank drive-up tellers? Air is used to push the tubes back and forth from the teller to the customer.
Air is also used to power drills, sanders, grinders, and the like at garages and car body shops.

182. Is shear thinning fluids and the linear visco-elastic fluids are the same?

Answer: No

183. What is the difference between the flight of a plane and a helicopter?
Answer: Both uses the pressure difference caused by air moving over the wings at different speeds to generate lift a plane by moving those wings in the direction of travel, a helicopter by spinning the "wings" around at high speed.

184. What is jet plane?
Answer: An aircraft powered by a jet engine.

185. How does a Coal Power station produce electricity?
Answer: A fossil fuel power plant (FFPP) (also known as steam electric power plant in the US, thermal power plant in Asia, or power station in the UK) is an energy conversion center designed on a large scale for continuous operation. Just as a battery converts relatively small amounts of chemical energy into electricity for temporary or intermittent use, the FFPP converts the energy stored in fossil fuels such as coal, oil, or natural gas successively into thermal energy, mechanical energy, and finally electric energy for continuous use and distribution across a wide geographic area.
Usually, the coal is utilized to convert water into steam in boilers (thermal energy). The steam is then used to drive steam turbines (mechanical energy). The turbine shaft is coupled to a generator shaft, which generates electricity.

186. What is the gas constant R for air?
Answer: Air gas constant is R_{air} = R/28.97=0.2869 (J/g K) = 286.9 (J/kg K)

187. Why is the suction pipe of vapour compression and refrigeration system insulated?
Answer: The vapour compression and refrigeration system is used to minimize sweating of the air around the pipe. When the air is exposed to a cool surface, its water vapour condenses.

188. What is the full form of WCB in ASTM?
Answer: American Society for Testing and Materials

189. What is the main purpose of an airspeed indicator in an aircraft?
Answer: The primary purpose of an airspeed indicator in an aircraft is to give the pilot some sense of how fast the aircraft is moving. A pitot tube, which is a forward pointing hollow tube that is mounted on the plane, is pressurized by the force of the air the plane encounters as it flies. This air pressure is compared to a static reference, and the difference is proportional to the airspeed. The faster the plane moves, the higher the pressure in the Pitot tube, and the greater the difference between that and the reference. All that will result in higher indicated air speed. There are some issues associated with the accuracy of the system (like when the plane is flying into a headwind), and information on. The pictures are informative, and a reader can pick up a handful of specialized terms relating to the device and the principles upon which it operates.

190. What was the spinning jenny?

Answer: The spinning jenny is a multi-spool spinning wheel. It was invented circa 1764 by James Hargreaves in Stanhill, near Blackburn, Lancashire in the north west of England (although Thomas Highs is another candidate identified as the inventor).

191. Who invented the four-stroke engine?
Answer: Nikolavs August Otto, German

192. What is meant by payload for aircraft?
Answer: The payload is the cargo that it carries.

193. How fast is a jet plane?
Answer: Fighter jet aircraft fly at speeds in the low mach numbers. Mach 1 is roughly 750 miles per hour, give or take, and military jets can fly at well over a thousand miles per hour. Back in the day (July of 1976), the SR-71 set the jet speed record. Radar clocked the thing at 2242 mph. Oh, and the record still stands. This highly sensitive aircraft project probably holds back data that would top this figure, but the information will not be coming out soon if it exists. The Blackbird is the answer to your prayers. As a caveat, there is a much faster jet called a scramjet, but it is unmanned. The NASA X-43A can hit Mach 9.6, which is about 7300 miles per hour.

194. What are the uses of windmill?
Answer: Windmills were traditionally used for processing grains, later they started to be used for electricity production as well. Windmills can also be used to pump water.

195. What is the world is hardest metal?
Answer: Rhenium diboribe

196. What is a machine shop?
Answer: A facility that uses machines to fabricate devices from stock raw materials or to modify mechanisms based upon provided specifications. Also known as "Back" Shops.
The common machines in a machine shop are a lathe, a drilling m/c, a miller, a shaper/planer, grinding machine and so on.

197. Why is thick steel wire stronger than thin steel wire?
Answer: The material will yield when stress reaches a critical value.
Stress = Load / Area
Thick steel wire is stronger than thin steel wire because there is more cross sectional area in the thick wire. Although the material's strength in load per unit area would be the same, the ultimate load that the wire can sustain would be more in the thick wire.
A simple way of looking at it is to imagine a thick wire as a number of thin wires stuck together. If a thin wire can support a mass of 1kg then 2 thin wires can support 2kg. A wire which is twice as thick (twice the cross sectional area) can also support 2kg.

198. What is the coefficient expansion of transformer oil?

Answer: Transformers can be filled with various types of Refined Mineral Oil. That coefficient is something you would find in specifications of the supplier.

199. What is a clock maker called?

Answer: A watchmaker or clockmaker is called a horologist.

200. What is 1810 stainless steel?

Answer: 1810 Stainless Steel is the European grade that is equivalent to AISI 304 Stainless Steel. It is the most common stainless steel going. Here is the rundown:
Fe, <0.08% C, 17.5-20% Cr, 8-11% Ni, <2% Mn, <1% Si, <0.045% P, <0.03% S

201. How can heat be applied to acrylic?

Answer: In most hardware stores, you can find what has called a Heat Gun. Hold the heat gun about 6 inches away from the acrylic you intend on bending say the bumper on a car. They are plastic these days, but will melt when you apply heat to them. Using a cold scraper, you can usually mold two pieces back together using the heat gun.

202. Did Elevation affect weight?

Answer: The acceleration of Gravity is less, as you get further away from the center of the earth. This would cause you to weigh less. The equation for the force of gravity is GMm/R^2. Where G is the universal Gravity Constant which is something like 6.67×10^{-11} and M is the mass of the earth, m is the mass of the weight being measured. R is the distance from the center of the earth that increases as you increase in altitude.

203. What is the Device that converts sound energy into mechanical energy?

Answer: Sound energy is mechanical energy. No devices are required to make a conversion.

204. What is a mechanism?

Answer: A mechanism is a system of moving parts that changes an input motion and force into a desired output motion and force.

205. Accelerometers measure acceleration. How do you measure the moment of inertia?

Answer: You calculate it using your moment of inertia equations corresponding to the geometry of the object. There is no simple device I do not think that you can buy to just measure it.

206. Who built the first rocket that Robert Goddard invented?

Answer: Goddard built his own early rockets.

207. What is a slined joint?

Answer: Cotter Pin

208. Why would some industries select pneumatics over hydraulic?
Answer: Actually, there is need for air reservoir in industrial pneumatics systems. Hydraulics can handle more powerful applications than pneumatics for the same overall dimensions, or can be more compact for the same power.

209. How do you convert 95 feet into square feet?
Answer: You cannot convert feet into square feet; they are two entirely different and incompatible units. Converting feet to square feet is like converting apples to oranges. A linear foot is a unit of distance. A square foot is a unit of area. If you have a rectangular surface, you can compute the area in square feet by multiplying the length in feet by the width in feet. For example, a rectangular patio that is 12 feet wide by 15 feet long has an area of 12×15 = 180 square feet.

210. What is difference between mill and mill-drill?
Answer: Today many manufactures are combining machines; a mill/drill is one of these. It is a combination of a drill and a mill, a mill removes stock from material (usually metal, but not limited just to metal, it depends on your application), you use fluted cutters such-as end mills. The drill aspect of the machine is just that it drills holes, with the proper speed for the right size drill.

211. What is a vee used for on the Vee Block?
Answer: It holds round/cylindrical material. It is usually used to judge the cylindricity of the cylindrical material.
Normally 90 degrees vee can be used for the marking of the cylindrical cross section surface, level checking.

212. How many litres did 1 cubic meter contains?
Answer: 1000 litre in a cubic meter

213. What is the weight of 1m3 of seawater?
Answer: $1 \text{ m}^3 = 1000$ lits
And 1 lit = 1 kg (approx)
Therefore $1 \text{ m}^3 = 1000$ kg (approx)

214. What are the advantages of powder metallurgy?
Answer: Power metallurgy is much faster production while holding closer sizes.

215. What are examples of a first class lever?
Answer: Examples of some first class levers are scissor, seesaw, hammer, and wrench.

216. How does a boat measure speed?
Answer: Knots one nautical mile per hour there are 1 1/8 of a nautical mile in a statute mile

Originally measured by throwing a piece of wood attached to a piece of rope over the back of a boat and counting how many knots went past in a given time.

These days it is more normally measured by:

a) Using a pitot static tube which measure difference in pressure and uses Bernoulli's equation to find the velocity.

b) Some form of propeller (technically impeller) which is suspended under the boat. The passing water turns, the speed of rotation is measured, and this gives you the speed of the boat.

c) Using ultrasonic to measure the speed, that small bubbles of air in the water passing under the boat go past.

217. What is a steam turbine diaphragm?

Answer: Steam turbine comprises of stages, number and size of the stages depends upon the break horsepower of the turbine.

The stage has set of moving and fixed blades. The moving blades are attached to the rotor while the stationary blades are called Diaphragm.

The diaphragm guides the steam to glide over the moving blades for producing rotary motion.

218. What is wrap-around?

Answer: The main purpose of a wrap-a-round is to make a straight line around a pipe to aid in cutting the pipe to its proper length. It is used mostly as a template or a straight edge.

219. What instrument in a car measures its speed?

Answer: The speedometer and speedometer cable tells the driver how fast the vehicle is going. What has called a Hall-Effect sensor is used. It uses the principle of magnetic inductance. When a magnetic flux passes through coils of wire, voltage is generated. To use this effect, a magnet is placed in the cars differential. The sensor then can tell when the magnet comes around by a spike in voltage. Since there is a constant amount the car moves with each differential rotation, and with the time between voltage spikes, you can easily divide to get the speed.

This is why changing your cars tires will affect your speedometer. Your car assumes that car moves a certain amount with each differential rotation. If you have larger tires on, then each differential rotation (and axle rotation) your car moves further, and you will move faster than indicated.

220. Can an airplane fly without a tail?

Answer: There are a number of aircraft, which are designed without tail assemblies. As for aircraft designed with a tail assembly that may lose it in flight, that's problematic for the flight crew, but there have been instances of portions of tail assemblies being lost due to structural failure or accident that have managed to successfully land.

221. Will going from a 3 tap to a 6 tap increase water pressure?

Answer: No, the pressure will be the same, you will get more volume only if your pumps can handle the gpm, to increase pressure you may need a booster pump or a single pump that is rated for your needs.

222. How does a hammer mill work?
Answer: Big hammers (not the kind you pound nails with) spin out with centrifugal force and beat whatever it is you are grinding into smaller pieces.

223. What are some mechanical laboratory apparatuses?
Answer: Welding machine

224. How does a modern submarine move?
Answer: Modern submarines move by using some motor to drive a propeller (called a screw). The big u-boats have nuclear reactors that heat water to make steam and have steam turbines that drive the screw through a big reduction gear. Many smaller submarines use a battery bank and electric motors with little propellers on them to move.

225. Is a diesel engines maximum rpm limited by the diesel burn rate?
Answer: Yes. In practice however, the maximum rpm is usually limited by the construction of the rotating assembly and the high-pressure injection parts.

226. What is the difference between "Stress and Strain"?
Answer: Stress is Load per unit Area. Strain is Change in Dimension (dL) Divided by Original Dimension (L), dL/L.

227. What is the Density of plastic?
Answer: Plastics are the general term for a wide range of synthetic or semi synthetic polymerization products. There are many different plastics; all have their own density! Most common plastics, however, have a density between .035 and .045 lb/cu in.

228. What are different types of gate valves?
Answer: There are a few different designs:
Parallel disk gate valves use two disks with a spring in between them, sliding into the seats. At low pressure, the spring forces the disks outward against the seats, sealing off the valve. At high pressure, all the sealing is accomplished by the downstream disk.
Wedge gate valves use a tapered disk that slides into two seats set at a slight, converging angle. The wedging action provides the sealing force between the disk and the seat.
Single disk gate valves are used where the flow is always one-way (like sluices on dams). There is some flexibility in the attachment of the disk to the stem, so the differential pressure on the disk pushes it against the seat, sealing it off.

229. Do you need to be an apprentice for mechanical engineering?

Answer: Most engineering jobs require at least a 4-year engineering degree. Since much of being an engineer is learned "on the job" it is good to have an internship or co-operative experience while in school, but it is not required for all engineering programs (some colleges do require it). You can get a job without an internship or co-op, but you should plan to "wow" them at your interview! It is always a good idea to get involved with engineering projects outside of the educational program, such as research or an engineering club. This shows employers that you have had some experience in a real working environment.

230. What type of paper when made into a paper boat floats longest?
Answer: I would use wax paper because of all the different kinds of paper it is the least absorbent.

231. Where do railway steam engines carry their water?
Answer: In the large tank on the front of the engine

232. Is there a metal that does not contain iron?
Answer: Most commercially available aluminium contains some other materials, but only accidental traces of iron, if any. For example, Copper, silver, and gold, same story.

233. How do concrete pumps work?
Answer: I wonder how concrete pumps work, liquid (fluid) concrete has a big viscosity, how the pump overcomes this big viscosity. I witnessed many times every couple seconds maybe 4-5 seconds the machine emits a loud thump. I think that sound come from pnomatic part of the machine.

234. What are the advantages of diesel engines?
Answer: The advantages of diesel engines are:
   a) Fuel economy
   b) Less maintenance
   c) Run cooler
   d) More power @ lower RPM's

235. How does an equal-arm balance work?
Answer: Two pans of equal balances are placed at the end of the beam, one at each end. A long pointer attached at right angles to the beam at the point of support. Zero on a scale indicates the beam is at rest.

236. Why do you need a keyway in the construction of a wall?
Answer: A keyway is found in a wall made of concrete where there is two separate pours, which abut themselves, otherwise referred to as a cold joint. It makes sense to insert a concave keyway continuously along the length and in the center of a section of the first wall section poured to receive an abutting concrete pour later. This inserted keyway creates an interlocking

style effect between the two abutting pours, which creates a higher quality connection than simply butting the two pours with nothing to interconnect them.

Typically, a keyway is found in a wall made of concrete where there is two separate pours, which abut themselves, otherwise referred to as a cold joint. It makes sense to insert a concave keyway continuously along the length and in the center of a section of the first wall section poured to receive an abutting concrete pour later.

This inserted keyway creates an interlocking style effect between the two abutting pours, which creates a higher quality connection than simply butting the two pours with nothing to interconnect them.

237. What is the history on the invention of paper?
Answer: Paper was invented in Egypt and was originally made from papyrus.

238. What kind of paper airplane flies the farthest?
Answer: Thickest paper has the greatest mass and therefore potential energy. Potential energy equals kinetic energy (speed). Speed equals lift. Lift equals a greater flight distance. Thicker planes fly farther.

239. What does a rolling offset look like?
Answer: Rolling offsets are used in the piping and sheet metal (ductwork) trades, a rolling offset changes the elevation and location of the piping or duct usually by using two fittings to offset around obstacles. Rolling offsets are used mostly when you are limited to the size of the fittings in order to change your elevation and location.

Rolling offsets are used in the piping and sheet metal (ductwork) trades, a rolling offset changes the elevation and location of the piping or duct usually by using two fittings to offset around obstacles. Rolling offsets are used mostly when you are limited to the size of the fittings in order to change your elevation and location

240. What is the difference between upstream and downstream in a refinery?
Answer: The total process of a the refining business starts at the oil field or gas field and runs all the way to the sending of processed hydrocarbon to a final user. Upstream applies to the operation of exploration, drilling, hydrocarbon production, and transmission via truck, rail or ship or pipeline to the refinery intake valve.

Downstream includes all work done at the refinery, distillation, cracking, reforming, blending storage, mixing and shipping. The case of heavy oil processing (oil sands etc.) and gas plant operation tend to cross the boundaries somewhat. Most are regarded as upstream operations even though downstream type operations are part of the processes. The production of chemical side products at gas plants (e.g. sulphur) is not generally segregated as a "Chemical Plant" operation.

Additional hydrocarbon production operations such as saddle plants, which remove a component from pipeline gas, are generally lumped with upstream.

341. What do you mean by the term thread size Please explain what does 0.125 Inch stand for in 0.125 Inch NPT thread?

Answer:
One thread per .125 inch or 8 threads per inch
One thread per .125 inch or 8 threads per inch

242. How many BTU per Hour equals 1 Ton of Air Conditioning?
Answer: Air conditioners are rated in three ways, by B.T.U.'s, Tons of Refrigeration, or by Horsepower. One Ton of Refrigeration removes the amount of heat needed to melt one ton of ice in 24 hours. One Ton of Refrigeration can remove 12,000 B.T.U.'s of heat in one hour. The B.T.U. is the amount of heat needed to raise 1 lb. of pure water 1 deg F.

243. How much water pressure will come from gravity feed tank, which is 10 metres high?
Answer: 956 pounds per square inch (psi)

244. What is the difference between hydraulic oil and engine oil?
Answer: Both the hydraulic and engine oils are made from base oils with additives mixed in. The additives used change the characteristics of the oils so that they function differently. Generally, hydraulic oils (final product including additives) are expected to have very low compressibility and very predictable friction and viscosity stability under pressure. Generally engine oils (Engine Lubrication Oils anyway) are intended to have high resistance to heat (degradation including chemical and viscosity due to heat) resistance to burning and resistance to absorption of fuels and chemical compounds produced during combustion. Both classes of oils are likely to have additives intended to provide detergency and to reduce foaming.
Base oils are most commonly petroleum oil bases due to cost, but other bases oil can be used including mineral oils (especially for hydraulic oils) and plant oils (especially for engine oils) and oils from animal sources.

245. What color are thermal oil piping lines?
Answer: Brownish and sort of grey

246. Which country has the world's most powerful tank?
Answer: Germany, USA, Russia, and France have the most powerful tanks. Germany's is probably the best after all the USA's man tank the Abrams Uses the German Gun.

247. Is Knowledge of properties of engineering materials is significant in pattern making why?
Answer: Yes, very important, wheels on car are spun on very hard steel die's not aluminium, the properties of the two metals are very different, one is soft, and the other is hard, (the wearing-out of the metals with extended use). In addition, how materials interact with each other as in coexistence and enhancement. Many things swell and shrink with heat and cold, linen is easier

to sew then fleece, patternmakers must take in all factors concerning the product they are working with.

248. How does hydraulics work?
Answer: A positive displacement pump (gear, vane, or piston pump) is driven by a prime mover (Electrical Motor or Engine) it sucks fluid from reservoir and delivers oil to system. During loading, a resistance to flow creates the pressure, which is utilized to do the work through cylinder for linear motion, or through hydraulic motor for rotary motion, Direction of flow is changed with help of direction control valve & system pressure is regulated by pressure control valve & flow is regulated by flow control valve.

249. What are examples of mechanisms?
Answer: Examples of mechanisms are the workings of a clock, a light switch, and a nail clipper.

250. What is the strongest pistol in the world?
Answer: Single shot pistol that shot a .50 BMG round is strong and could have been the strongest in the world.

251. What is a turboprop engine?
Answer: The combination of the words turbine and propeller in techno jargon will give you the word "turboprop". A turboprop engine is a turbojet (gas turbine) engine, which powers the propeller/s.
A conventional jet engine produces its thrust in large part due to the heated gasses escaping out the rear of the engine. While this is very useful for aircraft, which fly at high speeds and high altitudes, it is less desirable for aircraft, which are designed to fly at slower speeds and take off from smaller runways.
A turboprop engine is a jet engine, which converts the bulk of its thrust into rotational energy for powering a propeller. This allows jet engines, which are a high-rpm low-torque engine to be used in situations where low-rpm and high-torque are needed instead. The higher reliability and efficiency of a jet or turboprop engine as compared to an internal combustion engine makes them very desirable for aircraft designs, which, in the past, would have utilized internal combustion engines.

252. What is railroad track ballast?
Answer: That being said railroad track ballast is the cover for the subgrade. Ballast has several functions:
   a) It enables water to drain from the track
   b) It assists in helping control the thermal expansion of continues welded rail
   c) As a train passes the rails, it supports the train
   d) It distributes the weight of the train from the track structure to the sub grade
   e) Maintains a smooth running surface for the train to run on

As ballast becomes contaminated with other materials, it loses its ability to do some or all of these jobs.

253. How is a submarine able to submerge and surface?
Answer: Submarines have ballast tanks. The tanks can hold air, or they can be "vented" and water can flow into them and fill them completely. When the tanks are full of air, the submarine is buoyant and floats. When the main vents are opened, the tanks are flooded and the submarine submerges. When the submarine is submerged, wants to surface, air can be injected at high pressure into the ballast tanks to force out water, again make the submarine buoyant, and cause it to rise to the surface and float.

In practice, when a submarine surfaces, it doesn't use a lot of air from its high pressure air tanks to "blow the ballast tanks" because it takes a long time to pump up the high pressure air tanks again. What happens is that all the ballast tanks are given a "good shot" of high-pressure air (a few seconds), and then the planes are used in conjunction with the screw (which some call a propeller) to actually drive a submarine to the surface. Once there, something called a low-pressure blower system can be used to finish blowing the ballast tanks (while the high-pressure air compressors are running to pump the high-pressure air tanks back up).

254. Who built the Trans Alaska pipeline?
Answer: The pipe was constructed in six sections by five different contractors employing 21,000 people at the peak of work.

255. What are some examples of a periscope?
Answer: It is an optical instrument for viewing objects, which are above the level of direct sight; mostly used in submarines.

256. Is pipe round because it provides the least area to volume ratio?
Answer: More likely because it is easier to manufacture, much easier to put threads on, you do not have to worry about orientation when you put them together, and they have no weak spots created by corners.

My gutter pipes are rectangular because they do not stick out as far from the house as a circular one with the same area. However, they are low enough in the pressure they contain that they can be formed from sheet metal with a crimped seam. Making a water supply pipe, that way would be impossible.

In addition, it is easy to keep them aligned to the house. Running a rectangular water main under a street would be a major pain.

257. What is the difference between an electric motor and an electric generator?
Answer: There is no fundamental difference between an electric motor and an electric generator or dynamo. In normal use, all motors behave as generators, and all generators behave as motors. DC Motors act like generators because they use less electrical energy when allowed to spin fast.

DC generators act like motors because they become easier to spin when less electrical energy is drawn from their terminals.

For example, connect two small DC magnet motors together. Then if you spin the shaft of the first motor, the second motor's shaft will start spinning too. One acts as a DC generator, and the other acts as a DC motor. Alternatively, spin the second one's shaft, and the first one will start spinning.

Another example: If you connect a small DC motor to a small battery, then an electric current will appear in the motor's coils, and the motor starts spinning. However, if you spin the motor's shaft slightly faster than the normal speed, the direction of current in the circuit will reverse, and the battery starts taking in energy from the motor. The motor has become a generator, and it is recharging the battery.

258. How are the pneumatic system and the hydraulic system similar?
Answer: Pneumatics use gases such as air or nitrogen, hydraulics use oil or water, both systems use pressure to act on a specific application.

259. How can I see where pipes are behind the wall?
Answer: By using radio waves

260. Why do the radiators in your house click when you start the heating system?
Answer: They are clicking because they are heating up. The heat causes expansion and that is why you hear clicking.

261. How much is 1 KN in Kg?
Answer: A Newton, N, is a measure of force while a kilogram, kg, is a measure of mass. Therefore, any number of kN does not equal any number of kg.

This however is useful

$1 N = 1 kg.m/s^2$

$F = m \times a$

[F = force; m= mass; a=acceleration (g in case of acceleration due to gravity = $9.08665 m/s^2$)]

$m = F/g = 1/9.08665 = 0,11 Kg$

262. How does a super charger work in a car?
Answer: A supercharger is a used to increase the volume of air dragged into each cylinder per stroke.

In combustion engines, there are only really two ways to increase power, firstly increase the amount of fuel in the engine (either increase the displacement, or add more cylinders) or increase the amount of air in the engine (for a more effective explosion of the fuel)

Superchargers are powered by the engines crankshaft, which is connected usually by a belt drive. This rotational power turns a fan, which sucks air into the intake manifold similar in effect to turbochargers, with reduced lag. However, superchargers take more energy out the engine, so swings and roundabouts.

263. What is the Law of Thermodynamics?

Answer: There are actually 3 Laws of Thermodynamics (the actual number is debatable, but the number ranges from 3-5, depending on your adding skills).

264. How does hydraulic clutches work?

Answer: By using a non-compressible fluid, it acts like a solid push rod.

265. Will a steel cable become longer if it is heated while under load?

Answer: Even the heaviest steel cables stretch under load, whether they heat or not. Heating the cable will certainly elongate it. Friction caused by guides or pulleys can greatly increase the temperature of a cable under load.

You might consider the possibility of total failure if the heat is high enough and I am not sure but the cable might act in unpredictable ways when it fails. Applying heat to a steel cable under load is probably dangerous to your health and the health of anyone nearby.

When steel is heated, steel expands. It does not need to be under load. That is the reason that in the old days before advanced electronics and optics surveyors used chains rather than cables for measuring land. Had they used cable they would have gotten different measurements in summer and winter based on the difference in temperature and the coefficient of expansion of the material. The coefficient of expansion is a number that informs just how much a given material will expand or contract for a given change in temperature. To find the coefficient of expansion and other interesting information consult a materials handbook, available in better libraries or your local college of engineering. Another example of expanding steel and the necessity for dealing with this characteristic is the overlapping slip joint found on bridges that allows horizontal structural members to expand and contract without damaging the bridge.

266. What is the Congressional Space Medal of Honor?

Answer: The medal was created in 1969, designed by Congress for "any astronaut who in the performance of his duties has distinguished himself by exceptionally meritorious efforts and contributions to the welfare of the Nation and mankind.

267. Why will an ice cube placed in a cup of hot rotating water spin automatically faster than the water itself?

Answer: The center of a whirlpool is spinning the fastest. The cube is in the center so it is spinning equal or slower than the water directly under it but the water near the edge of the cup is spinning slower put something in the toilet, flush it, and observe how it spins faster as it reaches the center use something that will not clog the toilet.

268. Why is over-pressurizing an Air Conditioning system bad?

Answer: Overcharging a refrigeration or air conditioning system can result in an explosion. To avoid serious injury or death, never overcharge the system. Always use proper charging techniques. Limit charge amounts to those specified on the system equipment serial label or in the original equipment manufacturers service information.

Overcharging the system immerses the compressor motor, piston, connecting rods, and cylinders in liquid refrigerant. This creates a hydraulic block preventing the compressor from starting. The hydraulic block is also known as locked rotor.

Continued supply of electricity to the system causes heat to build in the compressor. This heat will eventually vaporize the refrigerant and rapidly increase system pressure. If, for any reason, the thermal protector fails to open the electrical circuit, system pressure can raise to high enough levels to cause a compressor-housing explosion.

269. What is the difference between blower and fan?
Answer: A fan can turn air, but a blower forces the air faster.

270. What is the date of first Titan launch from Cape Canaveral?
Answer: The Titan I was launched from Launch Pad 15 on February 6 1959.

271. What are the factors that can affect the Factor of safety selection?
Answer: The factor of safety is used in designing a machine component. Prior to selecting the correct factor of safety certain points must be taken into consideration such as:
1. The properties of the material used for the machine and the changes in its intrinsic properties over the time period of service.
2. The accuracy and authenticity of test results to the actual machine parts.
3. The applied load reliability.
4. The limit of stresses (localized).
5. The loss of property and life in case of failures.
6. The limit of initial stresses at the time period of manufacture.
7. The extent to which the assumptions can be simplified.

The factor of safety also depends on numerous other considerations such as the material, the method of manufacturing, the various types of stress, the part shapes etc.

272. What are the different types of fits? Explain?
Answer: On the basis of Indian standards fits can mainly be categorized into three groups:
1. **Clearance Fit**: These types of fits are characterized by the occurrence of a clearance between the two mating parts. The difference between the minimum size of the hole and the maximum size of the shaft is called the minimum clearance, the difference between the maximum size of the hole and the minimum size of the shaft is known as maximum clearance.
2. **Interference Fit:** In these types of fits the size of the mating parts are predefined so that interference between them always occurs. The tolerance zone of the hole is completely below the tolerance zone of the shaft.
3. **Transition Fit:** As the name suggests this type of fit has its mating parts sized limited to allow either clearance or interference. The tolerance zone of the hole and the shaft overlaps in case of such fits.

For a shaft designated as 40 H8/f7, calculate the tolerances.

Given: Shaft designation = 40 H8/f7

The shaft designation 40 H8/f 7 means that the basic size is 40 mm and the tolerance grade for the hole is 8 (i.e. I T 8) and for the shaft is 7 (i.e. I T 7).

Since 40 mm lies in the diameter steps of 30 to 50 mm, therefore the geometric mean diameter, D = Square root of (30 x 50) = 38.73 mm

We know that standard tolerance unit,

$$i = 0.45 \times \text{Cube root of } (D) + 0.001\ D$$
$$i = 0.45 \times 3.38 + 0.03873 = 1.559\ 73 \text{ or } 1.56 \text{ microns}$$
$$i = 1.56 \times 0.001 = 0.001\ 56 \text{ mm } ...(1 \text{ micron} = 0.001 \text{ mm})$$

The standard tolerance for the hole of grade 8 (IT8) = 25 i = 25 × 0.001 56 = 0.039 mm

The standard tolerance for the shaft of grade 7 (IT7) = 16 i = 16 × 0.001 56 = 0.025 mm

273. What is heat treatment and why is it done?

Answer: Heat treatment can be defined as a combination of processes or operations in which the heating and cooling of a metal or alloy is done in order to obtain desirable characteristics without changing the compositions. Some of the motives or purpose of heat treatment are as follows:

1. In order to improve the hardness of metals.
2. For the softening of the metal.
3. In order to improve the machinability of the metal.
4. To change the grain size.
5. To provide better resistance to heat, corrosion, wear etc.

Heat treatment is generally performed in the following ways:

1. Normalizing
2. Annealing
3. Spheroidising.
4. Hardening
5. Tempering
6. Surface or case hardening

274. What are the rules that must be kept in mind while designing castings?

Answer: Some of the points that must be kept in mind during the process of cast designing are as follows:

1. To avoid the concentration of stresses sharp corners and frequent use of fillets should be avoided.
2. Section thicknesses should be uniform as much as possible. For variations it must be done gradually.
3. Abrupt changes in the thickness should be avoided at all costs.
4. Simplicity is the key; the casting should be designed as simple as possible.
5. It is difficult to create true large spaces and henceforth large flat surfaces must be avoided.
6. Webs and ribs used for stiffening in castings should as minimal as possible.

7. Curved shapes can be used in order to improve the stress handling of the cast.

275. What are the points that should be kept in mind during forging design?
Answer: Some of the points that should be followed while forging design are:
1. A radial flow of grains or fibres must be achieved in the forged components.
2. The forged items such as drop and press forgings should have a parting line that should divide the forging into two equal halves.
3. The ribs in a forging should not be high or thin.
4. In order to avoid increased die wear the pockets and recesses in forgings should be minimum
5. In forgings the parting line of it should lie as far as possible in a single plane.
6. For ease of forging and easy removal of forgings the surfaces of the metal should contain sufficient drafts.

276. Describe briefly the different cold drawing processes.
Answer: Some of the important cold drawing processes are as follows:
1. **Bar and Rod Drawing:** In the case of bar drawing the hot drawn bars are at first pickled, washed and coated to prevent oxidation. Once this is done a draw bench is used for the process of cold drawing. In order to make an end possible to enter a drawing die the diameter of the rod is reduced by the swaging operation. This end is fastened by chains to the draw bench and the end is gripped by the jaws of the carriage. In this method a high surface finish and accuracy dimensionally is obtained. The products of this process can be used directly without any further machining.
2. **Wire Drawing:** Similar to the above process the bars are first pickled, washed and coated to prevent any oxidation. After this the rods are passed through several dies of decreasing diameter to provide a desired reduction in the size (diameter). The dies used for the reduction process is generally made up of carbide materials.
3. **Tube Drawing:** This type of drawing is very similar to the bar drawing process and in majority of cases it is accomplished by the use of a draw bench.

277. What are the different theories of failure under static load, explain briefly?
Answer: The main theories of failure of a member subjected to bi-axial stress are as follows:
1. **Maximum principal stress theory (Rankine's theory): This** theory states that failure occurs at a point in member where the maximum principal or normal stress in a bi-axial system reaches the maximum strength in a simple tension test.
2. **Maximum shear stress theory (Guest's or Tresca's theory):** This theory states that failure occurs when the biaxial stress reaches a value equal to the shear stress at yield point in a simple tension test.
3. **Maximum principal strain theory (Saint Venant theory):** This theory states that failure occurs when bi-axial stress reaches the limiting value of strain.

4. **Maximum strain energy theory (Haigh's theory):** This theory states that failure occurs when strain energy per unit volume of the stress system reaches the limiting strain energy point.
5. **Maximum distortion energy theory (Hencky and Von Mises theory):** This theory states that failure occurs when strain energy per unit volume reaches the limiting distortion energy.

278. What are the assumptions made in simple theory of bending?
Answer: The assumptions made in the theory of simple bending are:
1. The material of the beam is homogeneous this implies that it is uniform in density, strength and have isotropic properties meaning possessing same elastic property in all directions.
2. Even after bending the cross section of the beam remains constant.
3. During the initial stages the beam is straight and unstressed.
4. All the stresses in the beam are within the elastic limit of its material.
5. The layers of the beam are free to contract and expand longitudinally and laterally
6. On any cross section the perpendicular resultant force of the beam is zero.
7. Compared to the cross-sectional dimension of the beam the radius of curvature is very large.

279. Why stress is considered important in a shaft?
Answer: The following types of stresses are prevalent in shafts:
1. At the outermost surface of the shaft the max shear stress occurs on the cross-section of the shaft.
2. At the surface of the shaft on the longitudinal planes through the axis of the shaft the maximum longitudinal shear stress occurs.
3. At 45 degrees to the maximum shearing stress planes at the surface of the shafts the major principal stress occurs. It equals the max shear stress on the cross section of the shaft.
4. For certain materials where the tensile and compressive strengths are lower in measure as compared to the shear strength, then the shaft designing should be carried out for the lowest strengths.
5. All these stresses are of significance as they play a role in governing the failure of the shaft. All theses stresses get generated simultaneously and hence should be considered for designing purposes

280. What do you understand by the Hooke's Coupling what are its purposes?
Answer: The Hooke's coupling is used to connect two shafts whose axes intersect at a small angle. The two shafts are inclined at an angle and are constant. During motion it varies as the movement is transferred from one shaft to another. One of the major areas of application of this coupling is in gear boxes where the coupling is used to drive the rear wheels of trucks and other vehicles. In such usage scenarios two couplings are used each at the two ends of the coupling

shaft. They are also used to transfer power for multiple drilling machines. The Hooke`s coupling is also known as the Universal coupling. The torque transmitted by the shafts is given by:

$$T = (pie/16) \times t \times (d) \text{ cube}$$

Where T = torque, t = shear stress for the shaft material and d the diameter of the shaft.

281. What kind of materials should be used for shafts manufacturing?

Answer: Some of the qualities that should be present in materials for shafts are as follows:

1. The material should have a high index of strength.
2. Also it should have a high level of machinability.
3. The material should possess a low notch sensitivity factor.
4. The material must also have wear resistant properties.
5. Good heat treatment properties should also be present

The common material used to creates shafts of high strengths an alloy of steel like nickel is used. The shafts are manufactured by hot rolling processes and then the shaft is finished using drawing or grinding processes.

282. Why should a chain drive be used over a belt or rope driven drive? State pro`s and con`s?

Answer: The advantages of using a chain drives are:

1. In a chain drive no slip occurrence takes place.
2. The chains take less space as compared to rope or belts as they are made of metal and offer much strength.
3. The chain drives can be used at both short and long ranges and they offer a high level of transmission efficiency.
4. Chain drives can transmit more load and power as compared to belts.
5. A very high speed ratio can be maintained in one step of chain drives.
   Some of the cons of using a chain drive are:
6. The cost of producing chain drives is higher as compared to that of belts.
7. The chain drives must be serviced and maintained at regular intervals and henceforth their cost of ownership is high comparatively.

283. What are the different types of springs and explain them briefly?

Answer: Springs can be broadly classified into the following types:

1. **Helical Springs:** These springs as their name suggests are in coil form and are in the shape of helix. The primary purpose of such springs is to handle compressive and tensile loads. They can be further classified into two types: compression helical spring and tension helical spring each having their own unique areas of application.
2. **Conical and volute springs:** Both these spring types have specialized areas of usage where springs with adaptable rate according to the load is required. In case of conical springs they are wound so as to have a uniform pitch while on the other hand volute springs are wound in a slight manner of a paraboloid.

3. **Torsion Springs:** The characteristics of such springs are that they tend to wind up by the load. They can be either helical or spiral in shape. These types of springs are used in circuit breaker mechanisms.
4. **Leaf springs:** These types of springs are comprised of metal plates of different lengths held together with the help of bolts and clamps. Commonly seen being used as suspensions for vehicles.
5. **Disc Springs:** As the name suggests such types of springs are comprised of conical discs held together by a bolt or tube.
6. **Special Purpose Springs:** These springs are all together made of different materials such as air and water.

284. During the design of a friction clutch what are the considerations that should be made?

Answer: In order to design a friction clutch the following points must be kept in mind:
1. The material for the contact surfaces must be carefully selected.
2. For high speed devices to minimize the inertia load of the clutch, low weight moving parts must be selected.
3. The contact of the friction surfaces must be maintained at all the times without the application of any external forces.
4. Provisions for the facilitation of repairs must be there.
5. In order to increase safety the projecting parts of a clutch must be covered.
6. A provision to take up the wearing of the contact surfaces must be present.
7. Heat dissipaters to take away the heat from the point of contacting surfaces must be there.

285. What are the different types of brakes and explain them briefly?

Answer: Brakes can be classified on the basis of their medium used to brake, they are as follows:
1. **Hydraulic Brakes:** These brakes as their name suggest use a fluid medium to push or repel the brake pads for braking.
2. **Electric Brakes:** These brakes use electrical energy to deplete or create a braking force. Both the above types of breaks are used primarily for applications where a large amount of energy is to be transformed.
3. **Mechanical Brakes:** They can be further classified on the basis of the direction of their acting force.
A) **Radial Brakes:** As their names suggests the force that acts on the brakes is of radial direction. They can further be classified into internal and external blades.
B) **Axial Brakes:** In these types of brakes the braking force is acting in an axial direction as compared to radial brakes.

286. On what basis can sliding contact bearings be classified? Explain?

Answer: Sliding contact bearings can be classified on the basis of the thickness of the lubricating agent layer between the bearing and the journal. They can be classified as follows:

1. **Thick film bearings:** These type of bearings have their working surface separated by a layer of the lubricant. They are also known as hydrodynamic lubricated bearings.
2. **Thin film bearings:** In this type of bearings the surfaces are partially in direct contact with each other even after the presence of a lubricant. The other name for such type of bearings is boundary lubricated bearings.
3. **Zero Film Bearings:** These types of bearings as their name suggests have no lubricant present between the contact layers.
4. **Externally or hydrostatically pressurized lubricated bearings:** These bearings are able to without any relative motion support steady loads.

287. What are the basis on which the best material for Sliding Contact Bearings manufacturing?

Answer: Some of the important properties to lookout for in the material for sliding contact bearings are as follows:

1. **Compressive Strength:** In order to prevent the permanent deformation and intrusion of the bearing the material selected should be possess a high compressive strength to bear the max bearing pressure.
2. **Fatigue Strength:** the material selected for the bearing should be able to withstand loads without any surface fatigue cracks getting created. This is only possible if the material has a high level of fatigue strength.
3. **Comfortability:** The material should be able to adjust or accommodate bearing inaccuracies and deflections without much wear and heating.
4. **Embeddability:** The material should allow the embedding of small particles without affecting the material of the journal.
5. **Bondability:** The bearings may be created by bringing together (bonding) multiple layers of the material. Due to the above reason the bondability of the material should be sufficiently high.
6. **Thermal conductivity and corrosion resistance:** Thermal conductivity is an essential property for bearing materials as it can help in quickly dissipating the generated heat. Also the material should have a level of corrosion resistance against the lubricant.

288. Briefly explain the advantages of Cycloidal and Involute gears?

Answer: The advantages of the Cycloidal gears are as follows:

1. Having a wider flank as compared to Involute gears they are considered to have more strength and hence can withstand further load and stress.
2. The contact in case of cycloidal gears is between the concave surface and the convex flank. This results in less wear and tear.
3. No interference occurs in these types of gears.
   The advantages of Involute gears are as follows:

4. The primary advantage of involute gears is that it allows the changing of the centre distance of a pair without changing the velocity ratio.
5. The pressure angle remains constant from start to end teeth, this result in less wear and smooth running of the gears.
6. The involute gears are easier to manufacture as they can be generated in a single curve (the face and flank).

289. How can the reaction of support of a frame be evaluated?

Answer: Generally roller or hinged supports are used to support the frames. The conditions of equilibrium are used to determine the reaction support of a frame. The condition of equilibrium takes place when the sum of the horizontal and vertical forces sum equal to zero. The system must form a state of equilibrium even after considering the external loads and the reactions at the supports. For equilibrium to be prevalent in the system the following conditions are required to be in occurrence:

1. Summation of V = 0. This implies that the summation of all the forces in the vertical direction results to zero.
2. Summation of H = 0. This implies that the total of all the forces acting in horizontal direction is also zero.
3. Summation of M = 0. The sum of all the moment of forces around a point must be zero.

290. Explain in an orderly manner how the force in the member of a truss be detected using the method of joint.

Answer: The steps required to calculate the force are as follows:

1. The reaction at the support has to be first calculated.
2. Once the reaction is calculated the direction of force of the member is made to make it tensile. On getting the result to be negative the direction assumed is wrong and this implies the force being compressive in nature.
3. A joint need to be selected whose 2 members are not known. The lamis' theorem is used on the joint on which less than three forces are acting.
4. After the above process is complete the free body diagrams of the joint needs to be made. Since the system is in equilibrium the condition of Summation of V and H must result in zero.
5. After the above step the resolution of forces method needs to be used on the joint on which more than 4 forces are acting.

291. In order to derive the torsional formulas what are the assumptions taken?

Answer: The torsion equation is derived on the basis of following assumptions:

1. The shaft material is uniform, throughout the shaft.
2. Even after loading the shaft circular remains circular.
3. After the application of torques the plain section of a shaft remains plain.
4. Any twist that occurs in the shaft remains uniform and constant.

5. After the application of torque the distance between any two cross-sectional references remains constant.
6. The elastic limit value of a shaft is never exceeded even after the shear stress induced because of torque application.

292. What are Bevel Gears and what are its types?

Answer: Bevel gears are the type of gears in which the two shafts happen to intersect. The gear faces which are tooth bearing are conical in shape. They are generally mounted on shafts which are 90 degrees apart but they can be made to work at other angles as well. The bevel gears are classified into the following types on the basis of pitch surfaces and shaft angles:

1. **Mitre Gears:** These types of gears are similar to each other ie. they have the same pitch angles and contain the same number of teeth. The shaft axes intersect at 90 degrees angle.
2. **Angular bevel gears:** When two bevel gears connect at any angle apart from 90 degrees.
3. **Crown bevel gears:** When the two shaft axes intersect at an angle greater than 90 and one of the bevel gears have a pitch angle of 90 degrees they are known as crown bevel gears.
4. **Internal bevel gears:** In these type of gears the teeth on the gears is cut on the inside area of the pitch cone.

293. What are the different values that need to be determined in order to design a cylinder for an ICE?

Answer: The following values are needed to be determined:

1. Thickness of the cylinder wall: The cylinder walls in an engine are made witness to gas pressure and the side thrust of a piston. These results in two types of stresses: longitudinal and circumferential stress. Both the types of stresses are perpendicular to each other and hence it is aimed to reduce the resulting stress as much as possible.
2. Length and bore of the cylinder: The length of the cylinder and the length of the stroke is calculated on the basis of the formula: length of cylinder $L = 1.15$ times the length of the stroke (l). $L = 1.15(l)$
3. Cylinder flange and studs: The cylinders are always cast integral as a part of the upper crankcase or in some cases attached to it by means of nuts and bolts. The flange is integral to a cylinder and henceforth its thickness should be greater than that of the cylinder wall. The thickness of flange should generally be between 1.2t-1.4t where t is the cylinder thickness.
   The stud diameter is calculated by equating gas load (due to max pressure) to the grand total of all the resisting forces of the studs.

Piping Engineering

01. What are the types of compressors?
Answer: Positive Displacement, Centrifugal and Axial, rotary screw, rotary vane, rotary lobe, dynamic, liquid ring compressors.

02. What are types of compressor drives?
Answer: Electric motor, gas turbine, steam turbine and gas engine.

03. How Centrifugal compressors work?
Answer: High-speed impellers increase the kinetic energy of the gas, converting this energy into higher pressures in a divergent outlet passage called a diffuser. Large volume of gas at moderate pressure.

04. What are types of steam turbine and why are they popular?
Answer: Condensing and non-condensing, Popular because can convert large amounts of heat energy into mechanical work very efficiently.

05. Where gas turbine drive is used?
Answer: Desserts and offshore platforms where gas is available, for gas transmission, gas lift, liquid pumping, gas re-injection and process compressors.

06. What are the auxiliary equipments of compressor?
Answer: Lube oil consoles, Seal oil consoles, Surface condensers, Condensate pump, Air blowers, Inlet air filters, Wast heat system, compressor suction drum, knock out pot, Pulsation dampner, volume bottles, Inter and after coolers.

07. What are the types of seal oil system?
Answer: Gravity and pressurized.

08. What factors to be considered while designing compressor housing?
Answer: Operation, Maintenance, Climate conditions, Safety, Economics.

09. What are the compressor housing design points?
Answer: Floor elevation, building width, building elevation, hook centerline elevation.

10. What are the types of compressor cases?
Answer: Horizontal split case, Vertical split case.

11. How to located temperature and pressure instruments?

Answer: Temperature in liquid space, at down-comer side and pressure in vapour space, in area except down-comer sector.

12. What are necessary parts of inlet line of compressor?

Answer: Block Valve, Strainer, Break out flanges in both inlet and outlet to remove casing covers, Straightening vane in inlet line if not enough straight piece in inlet line available, PSV in interstage line and in discharge line before block valve.

13. What points to be considered for reciprocating compressor piping layout?

Answer: High pulsation, simple line as low to grade as possible for supporting, analog study, all branches close to line support and on top, Isolate line support from adjacent compressor or building foundations

14. What are the types of compressor shelters?

Answer: On ground with no shelter, Open sided structure with a roof, Curtain wall structure (Temperate climates), Open elevated installation, Elevated multi-compressor structure.

15. What are drum internals?

Answer: Demister pads, Baffles, Vortex breakers, Distribution piping.

16. What are drum elevation requirements?

Answer: NPSH, minimum clearance, common platforming, maintenance, operator access.

17. What are drum supports?

Answer: Skirt for large drums, legs, lugs, saddles for horizontal drums.

18. What are necessary nozzles for non-pressure vessel?

Answer: Inlet, outlet, vent, manhole, drain, overflow, agitator, temperature element, level instrument, and steamout connection.

19. What are necessary nozzles for pressure vessel?

Answer: Inlet, outlet, manhole, drain, pressure relief, agitator, level gauge, pressure gauge, temperature element, vent and for steamout.

20. What is preferred location for level instrument nozzles?

Answer: Away from the turbulence at the liquid outlet nozzle, although the vessel is provided with a vortex breaker, instrument should be set in the quiet zone of the vessel for example on the opposite side of the weir or baffle or near the vapour outlet end.

21. What is preferred location for process nozzles on drum?
Answer: Minimum from the tangent line.

22. What is preferred location for steam out nozzle on drum?
Answer: At the end opposite to the maintenance access.

23. What is preferred location for vent?
Answer: AT the top section of drum at the end opposite the steam out connection.

24. What is preferred location for pressure instrument nozzle on drum?
Answer: Must be anywhere in the vapour space, preferable at the top section of drum

25. What is preferred location for temperature instrument?
Answer: Must be in liquid space, preferably on the bottom section of drum.

26. What is preferred location for drain?
Answer: Must be located at the bottom section of drum.

27. What are the steps in selection of valve?
Answer: What to handle, liquid, gas or powder, fluid nature, function, construction material, disc type, stem type, how to operate, bonnet type, body ends, delivery time, cost, warranty.

28. What are functions of valves?
Answer: Isolation, regulation, non-return and special purposes.

29. What are isolating valves?
Answer: Gate, ball, plug, piston, diaphragm, butterfly, pinch.

30. What are regulation valves?
Answer: Globe, needle, butterfly, diaphragm, piston, pinch.

31. What are non-return valves?
Answer: check valve,

32. What are special valves?
Answer: multi-port, flush bottom, float, foot, pressure relief, breather.

33. What materials are used for construction of valves?
Answer: Cast iron, bronze, gun metal, carbon steel, stainless steel, alloy carbon steel, polypropylene and other plastics, special alloys.

34. What is trim?

Answer: Trim is composed of stem, seat surfaces, back seat bushing and other small internal parts that normally contact the surface fluid.

35. Which standard specifies trim numbers for valve?

Answer: API 600.

36. What are wetted parts of valve?

Answer: All parts that come in contact with surface fluid are called wetted parts.

37. What is wire drawing?

Answer: This term is used to indicate the premature erosion of the valve seat caused by excessive velocity between seat and seat disc, when valve is not closed tightly.

38. What is straight through valve?

Answer: Valve in which the closing operation of valve is achieved by 90degrees turn of the closing element.

39. What pressure tests are carried out on valves?

Answer: Shell-hydrostatic, seat-hydrostatic, seat-pneumatic

40. What are available valve operators?

Answer: Hand-lever, handwheel, chain operator, gear operator etc.

41. What is the full form of ASME?

Answer: American Society for Mechanical Engineers.

42. Which Piping code is used for Power piping and which code is used for Process Piping design?

Answer: Power Piping: ASME B 31.1

Process Piping: ASME B 31.3

43. What are the main differences between ASME B 31.1 and ASME B 31.3?

Answer: The main differences are listed below:
   a) Material allowable stresses are different in both codes.
   b) Stress increases due to occasional loads are different in each code.
   c) B 31.3 neglects torsion while calculating sustained stresses, but B 31.1 includes it.
   d) Sustained stress calculation is specific in B 31.1 but undefined for B 31.3.
   e) B 31.1 intensifies torsion but B31.3 does not intensify it.

44. How to calculate the basic allowable stress for a material?

Answer: The basic allowable stress is defined in respective code. For example as per B 31.3 the basic allowable stress for a material is the minimum of the following:

a) 1/3rd of tensile strength at design temperature.
b) 2/3rd of yield strength at design temperature.
c) 100% of average stress for a creep rate of 0.01% per 1000 hours.
d) 67% of average stress for rupture at the end of 100000 hours.
e) 80% of minimum stress for ruptures at the end of 100000 hours.
f) For austenitic stainless steel or nickel alloys the lower of yield strength and 90% of yield strength at temperature.
g) For structural grade materials 0.92 times of the lowest value of point a) to f)

45. What is the main difference between Constant and Variable Spring Hanger? When to use these hangers?

Answer: In Constant Spring hanger the load remains constant throughout its travel range. But In variable Spring hanger the load varies with displacement.

Spring hangers are used when thermal displacements are upwards and piping system is lifted off from the support position. Variable spring hanger is preferable as this is less costly.

Constant springs are used:

a) When thermal displacement exceeds 50 mm
b) When variability exceeds 25%
c) Sometimes when piping is connected to strain sensitive equipment like steam turbines, centrifugal compressors etc and it becomes very difficult to qualify nozzle loads by variable spring hangers, constant spring hangers can be used.

46. What do you mean by variability? What is the industry approved limit for variability?

Answer: Variability= (Hot Load-Cold load)/Hot load = (Spring Constant × displacement)/Hot load.

Limit for variability for variable spring hangers is 25%.

47. What are the major parameters you must address while making a Spring Datasheet?

Answer: Major parameters are: Spring TAG, Cold load/Installed load, Vertical and horizontal movement, Piping design temperature, Piping Material, Insulation thickness, Hydro-test load, Line number etc.

48. How to calculate the height of a Variable Spring hanger?

Answer: Select the height from vendor catalogue based on spring size and stiffness class.

For base mounted variable spring hanger the height is mentioned directly. It is the spring height.

For top mounted variable spring hangers ass spring height with turnbuckle length, clamp/lug length and rod length.

49. Can you select a proper Spring hanger if you do not make it program defined in your software?

What is the procedure?

Answer: In your system first decide the location where you want to install the spring. Then remove all nearby supports which are not taking load in thermal operating case. Now run the program and the sustained load on that support node is your hot load. The thermal movement in that location is your thermal movement for your spring. Now assume variability for your spring. So calculate

Spring constant = (Hot load × variability)/displacement. Now with spring constant and hot load enter any vendor catalogue to select spring inside the travel range.

50. What are the software available for performing piping stress analysis?

Answer: Caesar II, AutoPipe etc.

51. What are various temporary closures for lines?

Answer: Line blind valve, line blind, spectacle plate, double block and bleed, blind flanges replacing a removable spool.

52. Why horizontal displacement is specified in datasheet? What will you do if the angle due to displacement is more than 4 degree?

Answer: For bottom mounted springs it is mentioned to avoid large spring bending by frictional force and displacement. So that additional measures can be taken to lower frictional force by providing PTFE/graphite slide plate.

For top mounted spring hangers horizontal displacement is mentioned to check angularity of 4 degree to reduce transmission of horizontal force to piping systems as spring hangers are designed to take the vertical load only.

If angle becomes more than 4 degree due to large horizontal movement then install the spring hanger in a offset position so that after movement the angle becomes less than 4 degree.

53. Which spring will you select for your system: Spring with low stiffness or higher stiffness and why?

Answer: Springs with lower stiffness provides less load variation for same travel. So this spring is a better choice than a spring hanger with higher stiffness.

54. What do you mean by Stress? What are the types of Stresses that are generated in a Piping?

Answer: Whenever a force is applied to any object it applies a reaction force against the deformation by that force. That reaction force per unit area is the measure for the generated stress.

There has to be an external force to create stress. In a piping system there are various reasons for the generation of stress like Piping Weight, Internal and External pressure, Change in temperature, Seismic and Wind forces, PSV reaction force etc.

The stresses generated in a piping system are as follows:
  a) Axial Stresses
  b) Tangential or Hoop stress
  c) Longitudinal Stress.
  d) Radial Stress
  e) Expansion Stress
  f) Stress due to occasional events like Seismic and Wind effects.

55. What factors to consider for site selection?
Answer: District classification, Transportation facilities, Manpower availability, industrial infrastructure, community infrastructure, availability of raw water, effluent disposal, availability of power, availability of industrial gas, site size and nature, ecology and pollution.

56. Why Stress Analysis is required?
Answer: Ensure reliability and safety of working by
  a) Limiting Stresses (sustained, expansion, hydro-test, occasional) within code allowable.
  b) Limiting nozzle load and moments connected to equipment (Pump/Vessel/Heat Exchanger etc) within allowable values.
  c) Reducing damaging effects of dynamic loads.
  d) Avoiding leakage at joints.
  e) Limiting sagging and displacements within allowable limits.
  f) Avoiding high loads on supporting structures.

57. What is the difference between Stress and Pressure?
Answer: Stress is generated because of internal resistance force. Pressure is generated because of external force.
Pressure can be a cause to generate stress.

58. Where jacked screwed flange is used?
Answer: For spectacle discs, one flange is jacked screw flange. This flange has two jacked screws 180 degree apart which are used to create sufficient space between flange for easy removal and placement of line blind or spectacle blind.

59. What is double block and bleed?
Answer: Two valves with bleed ring in between with a bleed valve connected to the hole of bleed ring.

60. Where blind flange is used?
Answer: It is used with view to future expansion of the piping system, or for cleaning, inspection etc.

61. What are crude oil ranges?
Answer: Crude oil BP Range: 100F-1400F, lightest material: below 100F, Heavier materials-upto 800F, Residue above 800F.

62. What is batch shell process?
Answer: feed, heat, condense, heat more, condense, low quality.

63. What are types of towers?
Answer: Stripper, Vacuum tower, trayed, packed towers.

64. What is chimney tray?
Answer: It's a solid plate with central chimney section, used at draw-off sections of the tower.

65. What factors to consider while setting tower elevation?
Answer: NPSH, Operator access, Maintenance access, Minimum clearance, re-boiler type , common area, type of support, Tower dimensions, type of head, bottom outlet size, foundation details, minimum clearances.

66. How to located tower maintenance access nozzles?
Answer: At bottom, top and intermediate sections of tower, must not be at the down-comer section of tower and in front of internal piping.

67. How to located feed nozzle?
Answer: Must be oriented in specific area of tray by means of internal piping.

68. What are the steps in selection of valve?
Answer: What to handle, liquid, gas or powder, fluid nature, function, construction material, disc type, stem type, how to operate, bonnet type, body ends, delivery time, cost, warranty.

69. What are functions of valves?
Answer: Isolation, regulation, non-return and special purposes.

70. What are isolating valves?
Answer: Gate, ball, plug, piston, diaphragm, butterfly, pinch.

71. What are regulation valves?
Answer: Globe, needle, butterfly, diaphragm, piston, pinch.

72. What are non-return valves?
Answer: check valve.

73. What are special valves?
Answer: multi-port, flush bottom, float, foot, pressure relief, breather.

74. What materials are used for construction of valves?
Answer: Cast iron, bronze, gun metal, carbon steel, stainless steel, alloy carbon steel, polypropylene and other plastics, special alloys.

75. What is trim?
Answer: Trim is composed of stem, seat surfaces, back seat bushing and other small internal parts that normally contact the surface fluid.

76. Which standard specifies trim numbers for valve?
Answer: API 600.

77. What are wetted parts of valve?
Answer: All parts that come in contact with surface fluid are called wetted parts.

78. What is wire drawing?
Answer: This term is used to indicate the premature erosion of the valve seat caused by excessive velocity between seat and seat disc, when valve is not closed tightly.

79. What is straight through valve?
Answer: Valve in which the closing operation of valve is achieved by 90 degrees turn of the closing element.

80. What pressure tests are carried out on valves?
Answer: Shell-hydrostatic, seat-hydrostatic, seat-pneumatic

81. What are available valve operators?
Answer: Handlever, handwheel, chain operator, gear operator, powered operator likes electric motor, solenoid, pneumatic and hydraulic operators, Quick acting operators for non-rotary valves (handle lift).

82. What are ball valve body types?
Answer: Single piece, double piece, three piece, the short pattern, long pattern, sandwich and flush bottom design.

83. What are two types of ball valve?
Answer: Full port design and regular port design, according to type of seat, soft seat and metal seat.

84. Why ball valves are normally flanged?
Answer: Because of soft seat PTFE which can damage during welding.

85. What are butterfly valve types?
Answer: Double flange type, wafer lug type and wafer type.

86. What are types of check valve?
Answer: Lift check valves and swing check valves.

87. What are non-slam check valves?
Answer: Swing check valve, conventional check valve, wafer check valve, tilting disc check valve, piston check valve, stop check valve, ball check valve.

88. Where stop check valve is used?
Answer: In stem generation by multiple boilers, where a valve is inserted between each boiler and the main steam header. It can be optionally closed automatically or normally.

89. Where diaphragm valves are used?
Answer: Used for low pressure corrosive services as shut off valves.

90. What is Barstock Valve?
Answer: Any valve having a body machined from solid metal (barstock).
Usually needle or globe type.

91. What is BIBB Valve?
Answer: A small valve with turned down end, like a faucet.

92. What is Bleed Valve?
Answer: Small valve provided for drawing off liquid.

93. What is BlowDown Valve?
Answer: Refers to a plug type disc globe valve used for removing sludge and sedimentary matter from the bottom of boiler drums, vessels, driplegs etc.

94. What is Breather Valve?
Answer: A special self acting valve installed on storage tanks etc. to release vapour or gas on slight increase of internal pressure (in the region of ½ to 3 ounces per square inch).

95. What is Drip Valve?
Answer: A drain valve fitted to the bottom of a droplet to permit blowdown.

96. What is Flap Valve?

Answer: A non return valve having a hinged disc or rubber or leather flap used for low pressure lines.

97. What is Hose Valve?

Answer: A gate or globe valve having one of its ends externally threaded to one of the hose thread standards in use in the USA. These valves are used for vehicular and firewater connections.

98. What is Paper-Stock Valve?

Answer: A single disc single seat gate valve (Slide gate) with knife edged or notched disc used to regulate flow of paper slurry or other fibrous slurry.

99. What is Root Valve?

Answer: A valve used to isolate a pressure element or instrument from a line or vessel, or a valve placed at the beginning of a branch form the header.

100. What is Slurry valve?

Answer: A knife edge valve used to control flow of non-abrasive slurries.

101. What is Spiral sock valve?

Answer: A valve used to control flow of powders by means of a twistable fabric tube or sock.

102. What is Throttling valve?

Answer: Any valve used to closely regulate flow in the just-open position.

103. What is Vacuum breaker?

Answer: A special self-acting valve or any valve suitable for vacuum service, operated manually or automatically, installed to admit gas (usually atmospheric air) into a vacuum or low-pressure space. Such valves are installed on high points of piping or vessels to permit draining and sometimes to prevent siphoning.

104. What is Quick acting valve?

Answer: Any on/off valve rapidly operable, either by manual lever, spring or by piston, solenoid or lever with heat-fusible link releasing a weight which in falling operates the valve. Quick acting valves are desirable in lines conveying flammable liquids. Unsuitable for water or for liquid service in general without a cushioning device to protect piping from shock.

105. What is diverting valve?

Answer: This valve switch flow from one main line to two different outlets. WYE type and pneumatic control type with no moving part.

106. What is sampling valve?

Answer: Usually of needle or globe pattern, placed in branch line for the purpose of drawing all samples of process material thru the branch.

107. What are blow off valve?

Answer: It is a variety of globe valve confirming with boiler code requirements and specially designed for boiler blow-off service. WYE pattern and angle type, used to remove air and other gases from boilers etc.

108. What is relief valve?

Answer: Valve to relieve excess pressure in liquids in situations where full flow discharge is not required, when release of small volume of liquid would rapidly lower pressure.

109. What is safety valve?

Answer: Rapid opening (popping action) full flow valve for air and other gases.

110. What is foot valve?

Answer: Valve used to maintain a head of water on the suction side of sump pump, basically a lift check valve with integrated strainer.

111. What is float valve?

Answer: Used to control liquid level in tanks, operated by float, which rises with liquid level and opens the valve to control water level. It can also remove air from system, in which case, air flows out of system in valve open condition, but when water reaches valve, float inside valve raises to close the valve and stop flow of water. Used in drip legs.

112. What are flush bottom valves?

Answer: Special type of valves used to drain out the piping, reactors and vessels, attached on pad type nozzles.

113. What are types of flush bottom valves?

Answer: Valves with discs opening into the tank and valves with disks into the valve.

114. What are the uses of three-way valve?

Answer: Alternate connection of the two supply lines to a common delivery vise versa, isolating one safety valve, division of flow with isolation facility.

115. What are uses of four way valve?

Answer: Reversal of pump suction and delivery, By pass of strainer or meter, reversal of flow through filter, heat exchanger or dryer.

116. What is metal seated lubricated plug valve?
Answer: A plug valve with no plastic material, where grease is applied to contacting surfaces for easy operation.

117. What are three patterns of plug valve design?
Answer: Regular pattern, short pattern and ventury pattern.

118. What is regular pattern plug valve?
Answer: Rectangular port, area almost equal to pipe bore, smooth transition from round body to rectangular port, for minimum pressure loss.

119. What are short pattern plug valve?
Answer: Valves with face to face dimension of gate valve, as a alternative to gate valve.

120. What are ventury pattern plug valve?
Answer: Change of section through the body throat so graded to have ventury effect, minimum pressure loss.

121. What are inverted plug design valve?
Answer: Plug valve with taper portion up of plug. For 8" and higher size.

122. What is pressure balanced plug valve?
Answer: With holes in port top and bottom connecting two chambers on top and bottom of plug, to reduce turning effort.

123. What are Teflon sleeved plug valve?
Answer: PTFE sleeve between plug and body of valve, low turning effort, minimum friction, temperature limitation, anti static design possible.

124. What are permasil plug valve?
Answer: Plug valves with Teflon seat instead of sleeves, for on-off applications, can handle clean viscous and corrosive liquids, Graphite seat for high temperature applications. Drip tight shut off not possible.

125. What are eccentric plug valve?
Answer: Off center plug, corrosive and abrasive service, on off action, moves into and away from seat eliminating abrasive wear.

126. What is dimensional standard for plug valve?
Answer: API 599.

127. What is pinch valve?

Answer: Similar to diaphragm valve, with sleeves of rubber or PTFE, which get squeezed to control or stop the flow, Cast iron body, for very low service pressures like isolation of hose connections, manufacture standard.

128. What is needle valve?

Answer: Full pyramid disc, same design as globe valve, smaller sizes, sw or threaded, flow control, disc can be integral with stem, inside screw, borged or barstock body and bonnet, manufacturers standard.

129. How to install a globe valve?

Answer: Globe valve should be installed such that the flow is from the underside of the disk, Usually flow direction is marked on the globe valve.

130. What are globe valve port types?

Answer: Full port: More than 85% of bore size, Reducer port: One size less than the connected pipe.

131. What are globe valve disk types?

Answer: Flat faced type for positive shutoff, loose plug type for plug renewal or needle type for finer control.

132. What are characteristics of globe valve stem?

Answer: Always rising design, with disk nut at the lower end and handwheel at upper end.

133. What are types of globe valve?

Answer: Angle globe valve, plug type disc globe valve, WYE-body globe valve, composite disc globe valve, double disc globe valve.

134. What is angle globe valve?

Answer: Ends at 90 degree to save elbow, higher pressure drop.

135. Where plug type disc globe valve is used?

Answer: For severe regulating service with gritty liquids such as boiler feedwater and for blow off service.

136. Where WYE body globe valve is used?

Answer: In line ports with stem emerging at 45 degree, for erosive fluids due to smoother flow pattern.

137. What is double disc globe valve?

Answer: Has two discs bearing on separate seats spaced apart, on a single shaft, for low torque, used for control valves.

138. What are port types for gate valves?

Answer: Full port and reduced port. Default is reduced bore. Full port has to be specified in bom.

139. How to close a gate valve?

Answer: Turn the handwheel in clockwise direction.

140. What is lantern ring?

Answer: It's a collection point to drain off any hazardous seepage or as a point where lubricant can be injected; it is in the middle of packing rings.

141. What are types of gate valves?

Answer: Solid plane wedge, solid flexible wedge, split wedge, double disc parallels seats, double disc wedge, single disc single seat gate or slide, single disc parallel seats, plug gate valve.

142. What are the types of bonnets?

Answer: Bolted bonnet, bellow sealed bonnet, screwed on bonnet, union bonnets, A U-bolt and clamp type bonnet, breech-lock bonnet, pressure seal bonnet.

END

www.ingramcontent.com/pod-product-compliance
Lightning Source LLC
Chambersburg PA
CBHW081459170526

45166CB00008B/2479